Titanic Corrosion

Titanic Corrosion

Susai Rajendran
Gurmeet Singh

JENNY STANFORD
PUBLISHING

Published by

Jenny Stanford Publishing Pte. Ltd.
Level 34, Centennial Tower
3 Temasek Avenue
Singapore 039190

Email: editorial@jennystanford.com
Web: www.jennystanford.com

British Library Cataloguing-in-Publication Data
A catalogue record for this book is available from the British Library.

ISBN 978-981-4800-95-2 (Paperback)
ISBN 978-1-003-00449-3 (eBook)

Contents

Foreword

This book is excellently written and describes how marine corrosion could affect the deterioration of a historic ship like the *Titanic*. It is a superb reference book that every corrosionist and person involved  in the protection of national heritage monuments against corrosion should have as it may lead to prolonging the life of such monuments for future generations' knowledge. My gratitude and appreciation to the authors and their team for bringing such an great document to existence.

Dr. Abdulhameed Al-Hashem
Principal Research Scientist
Petroleum Research Center
Kuwait Institute for Scientific Research
P.O. Box 24885
Safat 13109, Kuwait

Preface

The word "titanic" is synonyms with colossal, gigantic, monumental, massive, enormous, towering, immense, vast, mammoth, huge, mighty, Herculean, Brobdingnagian . . . When one hears the word "titanic," he or she is reminded of the majestic ship *Titanic* and James Cameron's epic romance movie *Titanic*—in many cases the film first and the ship next because of the romance in the film, the songs, the camera work, vibrant sequences, the reality, and the ability to attract audiences from various walks of life, except a few orthodox oldies and religious-minded ones who give importance to spiritual values.

The colossal ship and the epic movie inspired the authors to study why the ship collapsed. Hundreds of reasons have been proposed, but the main reason seems to be bimetallic corrosion, also known as galvanic corrosion. The hull plates of the ship and the rivets that joined the hull plates were made of metals having different compositions. Having been in corrosive seawater for several months postlaunch, the hull plates and rivets became brittle and weak, and hence they easily broke when the ship collided with an iceberg on the silent, moonless night of April 14, 1912. Had the hull plates been malleable and ductile, they would have just bent in some places, which could have been repaired easily.

The collapse of this mighty ship has taught the massive mass many lessons. The danger of pride and faith in human achievement, the futility of wealth, and the danger of overconfidence are some of them. The courage and devotion to duty of the engineering crew (the real heroes of the *Titanic*) while the ship was sinking and the Good Samaritan, namely Captain Rostron of the *Carpathia*, who saved many human lives in the early hours of April 15, 1912, will be remembered by everyone forever.

The latest research on *Titanic* corrosion reveals that the ship is resting and reacting with bacteria and chloride ions, the corrosion culprit, on the Atlantic Ocean seabed. Once an elephantine ship, which served meat and wine to thousands of people in the first, second, and third classes and the duty-conscious, devoted workers and engineers, the *Titanic* has become food for the rust-eating bacterium *Halomonas titanicae.* The lustrous steel of the vast ship will one day become rust, reminding mankind of the everlasting lines from Ecclesiastes 3:20: *All go to the same place. All came from the dust and all return to the dust.*

The authors are thankful to their teachers, students, and friends, and to galvanic corrosion, who motivated them to write this book, *Titanic* Corrosion, and also to the publishers who constantly encouraged them to process the idea from a drafted manuscript to an excellent reference book.

Susai Rajendran
Gurmeet Singh
2020

Introduction

Corrosion is a natural and spontaneous process, just like death. The definition of corrosion; the causes, consequences, cost, and various forms of corrosion; and the methods of evaluating and controlling corrosion are presented in this section.

I.1 What Is Corrosion?

Corrosion is the gradual obliteration of materials through their chemical and/or electrochemical reaction with their environment.

Corrosion is a natural, spontaneous, and thermodynamically favorable process that converts a refined metal into a more chemically stable form, such as its oxide, hydroxide, or sulfide. Corrosion is the expression of the desire of the metal to go back to its original state.

I.2 Causes of Corrosion

For a metal to undergo corrosion, air (more precisely oxygen) and water are necessary. In the absence of any one of these, corrosion will not take place. This is illustrated in Fig. I.1.

Usually a pure metal will not undergo corrosion. However, when an impurity is introduced, the metal becomes nonhomogeneous in nature. Some sites act as anodes, and some sites act as cathodes. An anode has a tendency to lose electrons, and a cathode has a tendency to accept those electrons. For the electron transfer a medium, usually called an electrolyte, is needed (Fig. I.2). This event is called corrosion.

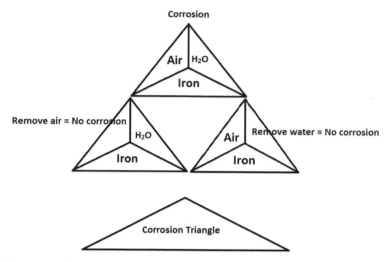

Figure I.1 The corrosion triangle.

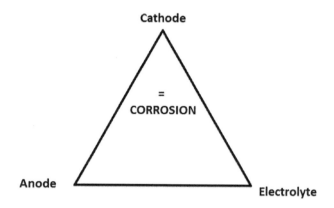

Figure I.2 The corrosiosn triangle.

I.3 Consequences of Corrosion

The consequences of corrosion are many and varied. Some of them are listed below.

- Escape of fluids
- Economic losses

- Loss of human lives
- Pollution problems—leakage of toxic gases
- Collapse of bridges
- Shutdown of factories

I.4   Cost of Corrosion

A study entitled "Corrosion Costs and Preventive Strategies in the United States" was conducted from 1999 to 2001 by CC Technologies Laboratories, Inc., with support from the US Federal Highway Administration and the National Association of Corrosion Engineers.

The study revealed that the cost can be classified into two types, direct and indirect.

A 1998 report revealed the annual direct and indirect costs of metallic corrosion in the United States and preventive strategies for optimum corrosion management. The total direct cost of corrosion is estimated at $276 billion per year, which is 3.1% of the 1998 US gross domestic product.

I.4.1   Direct Costs

The direct costs are divided into five major categories, given in Table I.1.

Table I.1   Direct costs of corrosion

Category	Amount
Infrastructure	$22.6 billion
Utilities	$47.9 billion
Transportation	$29.7 billion
Production and manufacturing	$17.6 billion
Government	$20.1 billion
Total	$137.9 billion

I.4.2 Indirect Costs

The indirect costs include:

- The cost of labor attributed to corrosion management activities
- The cost of the equipment required because of corrosion-related activities
- The loss of revenue due to disruption in the product supply
- The cost of loss of reliability

The grand total amounts to $275.5 billion.

Corrosion is so prevalent and takes so many forms that its occurrence and associated costs cannot be eliminated completely. However, it has been estimated that 25% to 30% of the annual corrosion costs in the United States could be saved if optimum corrosion management practices were employed.

The corrosion cost study clearly reveals that while technological advancements have provided many new ways to prevent corrosion, better corrosion management can be achieved using preventive strategies in nontechnical and technical areas. These strategies are identified as follows:

- Increase awareness of significant corrosion costs and potential cost savings.
- Change the misconception that nothing can be done about corrosion.
- Change policies, regulations, standards, and management practices to increase corrosion cost savings through sound corrosion management.
- Improve education and training of staff in the recognition of corrosion control.
- Implement advanced design practices for better corrosion management.
- Develop advanced life prediction and performance assessment methods.
- Improve corrosion technology through research, development, and implementation.

The United States continues to face critical challenges in the field of corrosion prevention and control, where aging equipment, new product formulations, environmental requirements, and strict budgets require corrosion control programs that are designed for specific situations and conducted by highly skilled professionals.

I.5   Forms of Corrosion

Depending on the environment and causes, corrosion has been classified into several types. However, the overlap of one type of corrosion with another cannot be ignored. Forms of corrosion are discussed in this section.

I.5.1   Uniform Corrosion

All metals in the atmospheric environment will undergo uniform corrosion. The driving forces of uniform corrosion are temperature and atmosphere. The corrosion cost of this type of corrosion is about 50% of the total cost of corrosion. Uniform corrosion can be controlled by painting and hot dip galvanization. Uniform corrosion is predictable, manageable, and often preventable.

I.5.2   Galvanic Corrosion

When two dissimilar metals are in contact with each other in the presence of an electrolyte, galvanic corrosion takes place. Galvanic coupling metals, for example, iron and copper, will undergo this type of corrosion. This is also known as two-metal corrosion. When iron is coupled with zinc, zinc will undergo corrosion. Iron will be protected. This principle is used in cathodic protection. When iron and copper are coupled, copper will be protected and iron will corrode because, in this case, iron has a tendency to lose electrons and copper has a tendency to accept electrons. Galvanic corrosion can be controlled by insulating the coupling materials. The collapse of the majestic ship RMS *Titanic*, when it collided with the iceberg, is attributed to galvanic corrosion. The hull plates and the rivets joining them, both under seawater, were of different compositions. Due to this, galvanic corrosion can be referred to as "*Titanic* corrosion" (Fig. I.3).

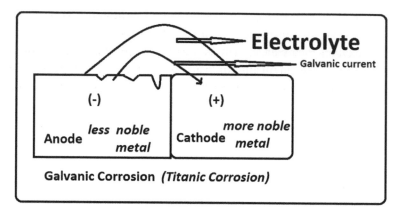

Figure I.3 Galvanic corrosion.

I.5.3 Intergranular Corrosion

All alloys undergo intergranular corrosion. The driving forces are the third-phase precipitate and temperature. Intergranular corrosion leads to the loss of strength and ductility. This type of corrosion can be controlled by heat treatment in manufacturing and welding during fabrication.

I.5.4 Crevice Corrosion

When there is a small gap between the two metals or between the metal and nonmetal in an electrolyte, crevice corrosion occurs (Fig. I.4). The gap may be <3.18 mm. If crevice corrosion prevails for a longer period, the corrosion problem will be severe. Crevice corrosion is an example of differential aeration corrosion. The metal part in the crevice part is an oxygen-poor area. It acts as the anode. So it undergoes corrosion. The other part of the metal acts as a cathode because it is rich in oxygen. Hence it will not undergo corrosion. Crevice corrosion can be controlled by proper designing, gasketing materials, and proper drainage designing.

I.5.5 Pitting Corrosion

Pitting corrosion (Fig. I.5) is another example of differential aeration corrosion. If a waterdrop or a dust particle is on the metal surface, the metal below the drop or particle is an oxygen-poor area and acts as an anode. It undergoes corrosion. The other part of the metal is an oxygen-rich area. It acts as a cathode and does not undergo corrosion. Localized pitting corrosion is initiated, and deep pits may result. We cannot see this corrosion per se but can see its ill effects. Pitting corrosion is considered as the second-biggest corrosion failure. Pitting corrosion will occur where surface irregularity exists and where chloride or bromide ions are present. Pitting corrosion can be controlled by surface quality control and proper welding practice.

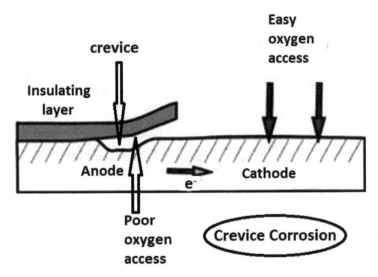

Figure I.4   Crevice corrosion.

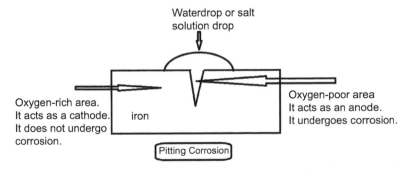

Figure I.5   Pitting corrosion.

I.5.6 Erosion Corrosion

Erosion corrosion is also known as tribo-corrosion. Carbon steel and stainless steel undergo erosion corrosion in flowing liquids containing abrasives. The driving force is the synergy effect of passive film breakdown by abrasive and localized corrosion. Severe attack may result in failure.

I.5.7 Stress Corrosion Cracking

Stress corrosion cracking (SCC) is the growth of crack formation in a corrosive environment. This will lead to sudden unexpected failure of normally ductile metals subjected to tensile stress, especially at elevated temperature. SCC is highly chemically specific in that certain alloys are likely to undergo SCC only when exposed to a small number of chemical environments.

Microbial corrosion (Fig. I.6) is the corrosion caused or promoted by microorganisms, usually chemoautotrophs. It can apply to both metals and nonmetallic materials. This corrosion is also known as bacterial corrosion. MIC can be controlled by addition of biocides, cleaning practice, and application of organic coatings. Sulphate-reducing bacteria, sulphate-oxidizing bacteria, and Fe-/Mn-oxidizing bacteria cause MIC. The *Titanic*, now at the bottom of the North Atlantic Ocean, is undergoing microbial-induced corrosion, resulting in the formation of rusticles.

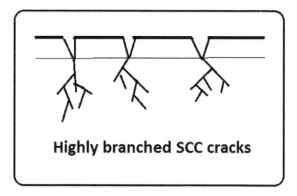

Highly branched SCC cracks

Figure I.6   Microbial-induced corrosion (MIC).

I.6 Methods Employed in a Corrosion Inhibition Study

To prevent corrosion of metals in solutions or in water (as in a cooling water system) corrosion inhibitors are added to the environment. In a corrosion inhibition study, the following methods are employed.

I.6.1 Weight Loss Method

Metal specimens of a suitable size are pickled with concentrated HCl, washed with water, and dried. Then the metal specimens are polished with emery papers of different grades to create a mirror finish, degreased with trichloroethylene or acetone, and dried. The weights of the metal specimens are found. Then they are immersed in test solutions, first in the absence and then in the presence of inhibitors. After a suitable immersion period, the metal specimens are taken out and washed with Clarke's solution and pure water. The weights of the dried metal specimens are found. From these data the inhibition efficiency is calculated from the relation

$$\text{inhibition efficiency (\%)} = [(w_1 - w_2)/w_1]100,$$

where w_1 = weight of the specimen in the absence of an inhibitor
w_2 = weight of the specimen in the presence of an inhibitor
Usually, three metal specimens will be used. The average of the three readings is taken.

I.6.2 Electrochemical Studies

Two electrochemical studies are used usually in order to know the mechanistic aspects of corrosion inhibition: polarization study and AC impedance spectra.

Electrochemical studies are carried out in a CHI-660 A work station model instrument. A three-electrode cell assembly is used. The working electrode is the metal specimen under investigation (mild steel, for example). Platinum foil is the counterelectrode. A saturated calomel electrode is the reference electrode (Fig. I.7).

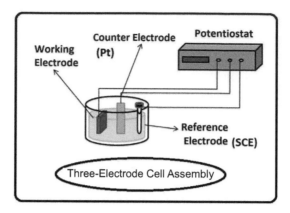

Figure I.7   A three-electrode cell assembly.

I.6.2.1   Polarization study

From the polarization study the following corrosion parameters are derived, from the Tafel plots: the corrosion potential (E_{corr}), the Tafel slopes (ba = anodic; bc = cathodic), the linear polarization resistance (LPR), and the corrosion current (I_{corr}).

The main principle is that when the corrosion resistance increases, the LPR value increases and the corrosion current decreases.

I.6.2.2   AC impedance spectra

From the AC impedance spectra, corrosion parameters such as charge transfer resistance (R_t), impedance value, double-layer capacitance (C_{dl}), and phase angle can be derived. For this purpose Nyquist plots and Bode plots are used. The experiments are carried out both in the absence and the presence of inhibitors. When corrosion resistance increases, the R_t value, the impedance value, and the phase angle value increase and the C_{dl} decreases.

I.6.3   Analysis of Protective Film

During the corrosion inhibition process, protective films are formed on the metal surface. The protective film can be analyzed from spectral studies.

Fluorescence spectra and Fourier-transform infrared spectra are used to confirm the formation of a protective film on the metal surface. The surface morphology can be analyzed by scanning electron microscopy, energy-dispersive X-ray spectroscopy, and atomic force microscopy.

From the analysis of the results of the above studies, a suitable mechanism for corrosion inhibition can be proposed.

I.7   Methods of Controlling Corrosion

There are many methods that are used to control the rate of corrosion. We can only control the rate of corrosion, not prevent it, just like we can only postpone death by treatments, not prevent it. We are born to die. Death is certain, just like corrosion of metals.

Some of the methods that are practiced are presented in the form of a chart (Fig. I.8). The best one is the use of nonmetallic materials. But it is impracticable. We cannot save ourselves by using plastic materials instead of metals and alloys. However, if we are able to produce strong plastic materials, then slowly, we can avoid using metallic materials.

The use of pure metals is also very difficult and expensive. In cooling water systems, as also in pipelines carrying oil, corrosion inhibitors can be used. Use of alloys will strengthen the metals and control corrosion. Painting is also widely used. Metallic coatings are used in various industries. Cathodic protection can be used to prevent the corrosion of hulls of ships.

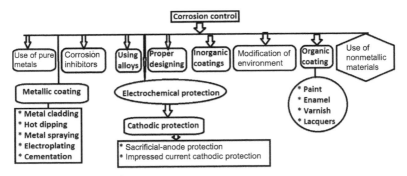

Figure I.8   Methods of controlling corrosion.

Suggested Readings

1. https://www.google.com/search?biw=1366&bih=657&tbm=isch&sa
   =1&ei=pakjXOmiOYbrQGJ2ZiICg&q=what+is+corrosion&oq=what+&
   gs_l=img.1.0.0i67l6j0l4.1332082.1337896..1339650...0.0..0.168.1987
   .0j16......0....1..gws-wiz-img.....0.FtdGUzYfUBU

2. https://www.google.com/search?biw=1366&bih=657&tbm=isch&sa
   =1&ei=464jXOLBEsPrvATTlI7QDQ&q=causes+of+corrosion&oq=caus
   es+of&gs_l=img.1.0.0i67j0l9.60902.66074..67654...0.0..0.500.2621.0j
   18j5-1......0....1..gws-wiz-img.....0.KKesTVrk66A

3. https://www.google.com/search?biw=1366&bih=657&tbm
   =isch&sa=1&ei=TK8jXNzcIYG6vwSVpKqACg&q=consequen
   ces+of+corrosion&oq=consequences+of+corrosion&gs_l=i
   mg.12...0.0..10505...0.0..0.0.0.......0......gws-wiz-img.ACkYWL6rZ1c

4. https://www.google.com/search?biw=1366&bih=657&tbm=isch&s
   a=1&ei=fK8jXLCVOYbVvgSawIPgAw&q=cost+of+corrosion&oq=cost
   +of+corrosion&gs_l=img.12...0.0..477...0.0..0.0.0.......0......gws-wiz-img.
   MfLTQJctC2o

5. https://www.google.com/search?biw=1366&bih=657&tbm=isch&sa
   =1&ei=r68jXOXXPIr-vgSFnIzwAQ&q=evaluation+of+corrosion

6. https://www.google.com/search?biw=1366&bih=657&tbm=isch&sa
   =1&ei=z68jXJ-OHsjGvgTbgpvQBw&q=forms+of+corrosion&oq=form
   s+of+corrosion&gs_l=img.12...0.0..66...0.0..0.0.0.......0......gws-wiz-img.
   ekCej6R7XI0

7. https://www.google.com/search?biw=1366&bih=657&tbm=isch&sa
   =1&ei=0a8jXIThBqOevQTi0YbYBQ&q=methods+of+prevention+of+c
   orrosion&oq=methods+of++of+corrosion&gs_l=img.1.0.0i7i30.33302
   .65383..70224...0.0..1.249.2016.0j15j1......0....1..gws-wiz-img.......0j0i7i
   5i30j0i8i7i30j0i8i30j0i24.QuUPB3bbv3A

Chapter 1

The RMS *Titanic*

"Titanic" means gigantic. The Royal Mail Ship (RMS) *Titanic* (subsequently referred to as the *Titanic*) (Fig. 1.1) was the world's largest passenger ship when she entered service. She was as long as three football fields (882 feet, or 269 m, in length). Her height was 32 m (104 feet). The volume of space within her hull and the enclosed space above the deck totalled 2831.68 liters (46,328 gross register tons), and she displaced 52,310 tons. [1]. She was one of the first ships to have a telephone system and electric lights in all the rooms. She had four elevators, a heated swimming pool, two barber shops, two libraries, a kennel for first-class dogs, a Turkish bath, a squash court, and a gymnasium; she even had her own on-board newspaper, *The Atlantic Daily Bulletin*.

The *Titanic* carried 16 wooden lifeboats and 4 collapsible boats were carried, enough to accommodate 1178 people.

On the night of April 14, 1912, on her maiden voyage, the *Titanic* collided with an iceberg and sank in the early hours of April 15, 1912. Over 1500 people died. The remains of the *Titanic*'s wreck sit on the seabed more than 12,415 feet (3784 m) beneath the surface, resting and reacting with seawater and microorganisms.

Titanic Corrosion
Susai Rajendran and Gurmeet Singh
Copyright © 2020 Jenny Stanford Publishing Pte. Ltd.
ISBN 978-981-4800-95-2 (Paperback), 978-1-003-00449-3 (eBook)
www.jennystanford.com

Figure 1.1   The *Titanic* [1].

1.1   Special Features of the *Titanic*

1.1.1   Power

There were three main engines in the *Titanic*. Each engine drove a propeller efficiently (Fig. 1.2) [1]. The total output of the three engines was 46,000 hp. When the three engines worked together, the motive power increased and the fuel usage was reduced. There were four 400 kW steam-driven electric generators near the turbine engine. They provided electrical power to the *Titanic* until the last few minutes before the ship sank [1].

1.1.2   Technology

1.1.2.1   Watertight compartments and funnels

There were 16 water tight compartments in the *Titanic*, as in any other Olympic-class ship, 15 bulkheads, and 11 vertically closing watertight doors to seal off the 16 compartments in any emergency. The exposed decking of the *Titanic* was made of pine and teak. Four funnels were standing above the decks. Of them, only three were functional, while the fourth one was a dummy, installed for aesthetic purposes and kitchen ventilation) [1].

Figure 1.2 Rudder with central and port wing propellers of the *Titanic*. For scale, note the man at the bottom of the photograph [1].

1.1.2.2 Rudder and steering engines

The *Titanic*'s rudder was so large was 23.98 m high and 4.65 m long and weighed more than 100 tons. It was so heavy that only steering engines could move it. Two steam-powered steering engines were used alternatively to move the rudder. The ship also had five anchors:

one port, one starboard, one in the centerline, and two kedging anchors [1].

1.1.2.3   Water, ventilation, and heating

The *Titanic* was provided with a well-planned and properly regulated network of pipes and valves. Heating and pumping water to all parts of the vessel was done through this system. If necessary, seawater was also used, but it would result in formation of scales, which required additional heat energy. Electric heaters were used to supply hot air [1].

1.1.2.4   Radio communication

The Marconi International Marine Communication Company looked after the *Titanic's* wireless telegraphy. The company provided two operators, namely Jack Phillips and Harold Bride, to take care of the 24 h service. Their main work was sending and receiving the passengers' messages. They also took care of weather reports and ice warnings.

The soundproof radio room, the "Silent Room," was located in the officers' quarters. It was provided with a motor generator and a transmitter. The *Titanic* had a state-of-the-art 5 kW rotary spark gap transmitter, operating under the radio callsign MGY, with communication being conducted in Morse code [1]. The transmitter was very powerful and was able to broadcast over a radius of 563 km. This rotary spark gap transmitter the *Titanic* its distinctive musical tone readily distinguished from other signals [1]. A ship-length T-antenna was used for transmitting and receiving messages.

1.1.3   Lifeboats

The *Titanic* was equipped with 20 lifeboats—14 standard wooden Harland and Wolff lifeboats with a capacity of 65 people each and 4 Englehardt "collapsible" (wooden bottom, collapsible canvas sides) lifeboats (identified as A to D) with a capacity of 47 people each [1]. Each boat carried a spare life belt, blankets, food, and water. In addition, the *Titanic* had 2 emergency cutters with a capacity of 40 people each. These boats could accommodate about 1178 people, which was only one-third of the *Titanic's* total capacity. However , British vessels over 10,000 tons could carry only 16 lifeboats each, with a capacity of 990 people, as per the Board of Trade's regulations (Fig. 1.3) [1].

Figure 1.3   A lifeboat [1].

1.2   History and General Characteristics of the *Titanic*

The history and general characteristics of the *Titanic* are given in Tables 1.1 and 1.2.

Table 1.1   History of the *Titanic* [1]

History	
	United Kingdom
Name:	RMS *Titanic*
Owner:	White Star Line
Port of registry:	Liverpool, UK
Route:	Southampton to New York City

(Continued)

Table 1.1 (*Continued*)

History	
Ordered:	September 17, 1908
Builder:	Harland and Wolff, Belfast
Cost:	7.5 million (US dollars)
Yard number:	401
Laid down:	March 31, 1909
Completed:	May 31, 1911
Launched:	April 2, 1912
Maiden voyage:	April 10, 1912
In service:	April 10–15, 1912
Identification:	Radio call sign "MGY"
Fate:	Hit an iceberg at 11:40 p.m. (ship's time) on April 14, 1912, on her maiden voyage and sank 2 h 40 min. later
Status:	Wreck

Table 1.2 General characteristics of the *Titanic* [1]

General characteristics	
Class and type:	Olympic-class ocean liner
Tonnage:	46,328 GRT
Displacement:	52,310 tons
Length:	882 ft 9 in (269.1 m)
Beam:	92 ft 6 in (28.2 m)
Height:	175 ft (53.3 m) (keel to top of funnels)
Draught:	34 ft 7 in (10.5 m)
Depth:	64 ft 6 in (19.7 m)
Decks:	9 (A–G)
Installed power:	24 double-ended and five single-ended boilers feeding 2 reciprocating steam engines for the wing propellers and 1 low-pressure turbine for the center propeller [1]; output: 46,000 hp
Propulsion:	Two three-blade wing propellers and one four-blade center propeller
Speed:	Cruising: 21 kn (39 km/h; 24 mph). Max.: 24 kn (44 km/h; 28 mph)

General characteristics	
Capacity:	Passengers: 2435; crew: 892. Total: 3327 (or 3547 according to other sources)
Notes:	Lifeboats: 20 (sufficient for 1178 people)

1.3   Maiden Voyage

On noon on Wednesday, April 10, 1912, the *Titanic*'s began her maiden voyage, between Southampton and New York. During her journey, she halted at Southampton, Cherbourg, and Queenstown. On Thursday, April 11, 11:30 a.m., she arrived at Cork Harbour. At 1:30 p.m., she departed on her westward journey across the Atlantic Ocean (Fig. 1.4) [1].

Figure 1.4   The *Titanic* at Southampton docks, prior to departure [1].

On April 14, about 600 km south of Newfoundland, the *Titanic* hit an iceberg at 11:40 p.m. Because of this collision, her hull plates buckled inward along her starboard (right) side and opened 6 of her 16 watertight compartments to the sea. Some people, mainly women and children, were evacuated in partially loaded lifeboats. At 2:20 a.m, on April 15, the *Titanic* sank. Two hours later, the Cunard liner RMS *Carpathia* came to the accident site and saved around 705 survivors.

In 1985, more than 70 years after the disaster, the wreck of the *Titanic* was discovered at a depth of 3784 m (12,415 feet) on the Atlantic Ocean seabed. The ship is split in two and is gradually disintegrating [1].

The *Titanic* is remembered through many works of popular culture, such as books, films, folk songs, exhibits, and memorials [1].

Ever since this disaster, a number of movies have been made on the tragedy:

- *Saved from the Titanic* (1912)
- *In Nacht und Eis* (1912)
- *Atlantic* (1929)
- *Titanic* (1943)
- *Titanic* (1953)
- *A Night to Remember* (1958)
- *S.O.S. Titanic* (1979)
- *Raise the Titanic* (1980)
- *The Chambermaid on the Titanic* (1998)
- *Titanic: The Legend Goes On* (2001)
- *Titanic II* (2010)
- *The Unsinkable Molly Brown* (1964)
- *Titanic* (1997)

The chapters of this book discuss the various aspects of galvanic corrosion, such as causes, consequences, methods of control, and case studies. They also report research on the causes of corrosion of the *Titanic*, including microbiologically influenced corrosion (MIC) and metallurgical failure.

Reference

1. https://en.wikipedia.org/wiki/RMS_Titanic (https://en.wikipedia. org/wiki/RMS_Titanic)

Chapter 2

Galvanic Corrosion of the *Titanic*

The Royal Mail Ship (RMS) *Titanic* was on her maiden voyage from Southampton to New York City, when on April 15, 1912, early in the morning, at 2:20 a.m., she sank after colliding with iceberg the night before [1]. More than 1500 people lost their lives that night (Table 2.1) [1].

Table 2.1 Survivors, victims, and statistics [1]

Category	Number aboard	Number of survivors	Percentage survived	Number lost	Percentage lost
First class	329	199	60.5%	130	39.5%
Second class	285	119	41.7%	166	58.3%
Third class	710	174	24.5%	536	75.5%
Crew	899	214	23.8%	685	76.2%
Total	**2223**	**706**	**31.8%**	**1517**	**68.2%**

Everyone wonders whether the Royal Mail Ship (RMS) *Titanic* became famous because of James Cameron's movie *Titanic* (1997)

Titanic Corrosion
Susai Rajendran and Gurmeet Singh
Copyright © 2020 Jenny Stanford Publishing Pte. Ltd.
ISBN 978-981-4800-95-2 (Paperback), 978-1-003-00449-3 (eBook)
www.jennystanford.com

or whether the movie became famous because of the ship *Titanic*. For us, and other scientists like us, the *Titanic* is famous because she sank due to galvanic corrosion, also known as bimetallic corrosion.

2.1   Construction, Launch, and Fitting-Out of the *Titanic*

The *Titanic* and her sister ship the *Olympic* were constructed on Queen's Island, now known as the Titanic Quarter, in Belfast Harbour. Their construction was facilitated by the Arrol Gantry, which was 69 m (228 feet) high, 82 m (270 feet) wide, and 260 m (840 feet) long, weighed more than 6000 tons, and could accommodate several mobile cranes (Fig. 2.1) [2].

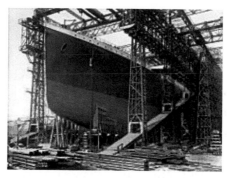

Figure 2.1   Construction of the *Titanic* in the gantry, 1909–1911 [1].

On March 31, 1909, the *Titanic*'s keel was laid down. The ship was built over a period of 26 months. She was framed as a huge floating box girder, the keel acted as the backbone, and the frames of the hull formed the ribs [2]. The 2000 hull plates used in the construction of the *Titanic* were single pieces of rolled steel plate, up to 1.8 m wide, 9.1 m long, and 3.8 cm thick. The plates were laid in an overlapping fashion. Commercial oxy-fuel and electric arc welding techniques were in their initial stages during that period, so over 3 million iron and steel rivets, weighing more than 1200 tons, were used to hold the hull together [2].

Decades after the disaster, some material scientists concluded that the steel plates became brittle because of the cold, exacerbating the impact damage and hastening the ship's sinking. Considering the standards of that time, the steel plates' were of good quality, not faulty, but were inferior compared to those used now because of advances in steelmaking metallurgy [2].

Construction of the *Titanic* was difficult and dangerous. During the *Titanic*'s construction, "246 injuries were recorded, 28 of them 'severe,' such as arms severed by machines or legs crushed under falling pieces of steel. Six people died on the ship herself while she was being constructed and fitted out, and another two died in the shipyard workshops and sheds. Just before the launch a worker was killed when a piece of wood fell on him" [2].

Finally, on May 31, 1911, at 12:15 p.m., the *Titanic* was launched (Figs. 2.2 and 2.3) [2].

Figure 2.2   Launch of the *Titanic*, 1911 (unfinished superstructure) [1].

Figure 2.3   Fitting-out of the *Titanic*, 1911–1912 [1].

2.2 Collision with the Iceberg and Collapse of the *Titanic*

The Titanic was traveling at 22.5 knots. This was just 0.5 knots below her maximum speed capability. The ship was clearly traveling too fast for the conditions, despite the fact that she received a series of warnings of drifting ice in the area of the Grand Banks of Newfoundland from other ships.

The captain ordered the engines to be reversed, which sealed the Titanic's fate. If the Titanic had sailed on ahead and then turned, in all probability she would have not hit the iceberg. The collision cut a 220–245-foot gash into the hull of the Titanic. However, recent evidence shows that an opening the size of a refrigerator allowed water to enter the ship. In addition, the 16 watertight compartments in the hull were not actually watertight but open at the top, worsening the situation. The ship could have stayed afloat with 4 compartments flooded, but 6 compartments being flooded doomed her.

2.3 Galvanic Corrosion Sank the "Unsinkable" *Titanic*

Galvanic exchange is considered the real reason for the sinking of the *Titanic* sank. Briefly, 3 million wrought iron rivets were used to join the hull plates. The rivets and hull plates had different compositions. During her construction, the *Titanic* sat in seawater, which completed an electric circuit. So the rivets corroded faster than the hull plates, resulting in the collapse of the ship when she collided with the iceberg [3]. In the collision (Fig. 2.4), the already weakened rivets popped. A 1-inch-wide opening was created between the hull plates, which was sufficient for seawater to gush into the ship and sink her [3]. "Video images of the wreckage show a narrow opening in the unburied part of the bow, as well as preferential corrosion of the rivets in some areas" [3].

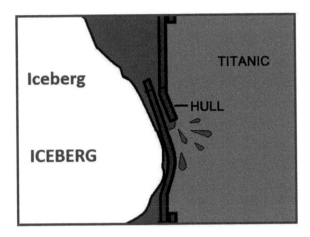

Figure 2.4 The initial damage caused to the *Titanic* by the iceberg. Image: https://en.wikipedia.org/wiki/Sinking_of_the_RMS_Titanic (https://en.wikipedia.org/wiki/Sinking_of_the_RMS_Titanic).

According to Robert Baboian, the retired director of Texas Instruments' corrosion lab, the *Titanic* was a victim of rust. The dissimilar metals of the hull and rivets, bathed in electrically conductive seawater, might have created a circuit that slowly flecked away and weakened the rivets. "One of the last photos taken before the ship's maiden voyage shows a pattern that suggests the rivets were rusting faster than the hull plates," says Baboian.

With regard to the narrow opening in the unburied part of the bow and the preferential corrosion of the rivets, Baboian says, "It is about at the level where the iceberg would have struck, and it is right where rivet popping could occur. I think that caused the *Titanic* to sink." The following information supports his theory that corrosion contributed to the sinking of the *Titanic* [1]:

- She was launched to be fitted out at the dockside almost a year before her maiden voyage (Figs. 2.2 and 2.3).
- During the year in seawater, there is a strong possibility that stray currents from the direct current (DC) equipment caused accelerated corrosion.
- The rivet iron was different from the hull plate iron by design. The rivet iron needed to be malleable and therefore consisted of a higher level of slag inclusions.

- During the year at the dockside, corrosiveness of the rivet iron could have been higher than the hull plate iron's. Galvanic corrosion [4] of the rivet iron is likely during that period.
- Photos and videos taken by Robert Ballard during his 1985 and 1986 expeditions to the *Titanic* show preferential corrosion of the rivets in some areas.
- Inspection of the big piece that was retrieved from the bottom of the ocean also shows this same preferential corrosion in some areas. All of this evidence points toward weakening of the hull–rivet structure, enhancing the rivet-popping mechanism that caused the *Titanic* to sink.

References

1. Rajendran, S. (2016). Titanic corrosion (invited talk), *International Journal of Nano Corrosion Science and Engineering*. https://www.researchgate.net/publication/308961615_Invited_ Talk_-Titanic_corrosion_Invited_Talk_-Titanic_corrosion_Invited_Talk_- Titanic_corrosion

2. https://en.wikipedia.org/wiki/RMS_Titanic (https://en.wikipedia.org/wiki/RMS_Titanic)

3. National Oceanic and Atmospheric Administration (http://oceanexplorer.noaa.gov). https://oceanexplorer.noaa.gov/ edu/lessonplans/Titanic04.Galvanic.pdf

4. https://en.wikipedia.org/wiki/Galvanic_corrosion (https://en.wikipedia.org/wiki/Galvanic_corrosion)

Chapter 3

Galvanic Corrosion

When two dissimilar metals are in contact with each other in the presence of an electrolyte, electrolysis takes place and one metal corrodes in preference to the other. One well-known example of electrolysis is the reaction between zinc and iron. Zinc atoms ionize because zinc is more electronegative and is oxidized and corrodes.

$$Zn\ (s) \rightarrow Zn^{2+}\ (aq.) + 2e\ (oxidation)$$

When dissimilar metals having different electrode potentials come in contact with each other in the presence of an electrolyte, one metal acts as the anode and the other as the cathode. The anode undergoes corrosion. For example, when zinc and iron are in contact with each other in the presence of an electrolyte, such as sodium chloride solution, zinc acts as the anode and undergoes corrosion in preference to iron [1]. "The presence of an electrolyte and an electrical conducting path between the metals is essential for galvanic corrosion to occur. The electrolyte offers a means for ion migration, whereby ions move to prevent charge build-up that would otherwise stop the reaction" [1].

Various aspects of galvanic corrosion, namely causes, consequences, prevention, and case studies, are discussed in this chapter.

Titanic Corrosion
Susai Rajendran and Gurmeet Singh
Copyright © 2020 Jenny Stanford Publishing Pte. Ltd.
ISBN 978-981-4800-95-2 (Paperback), 978-1-003-00449-3 (eBook)
www.jennystanford.com

3.1 Galvanic Series

All metals can be classified into a galvanic series representing the electrical potential they develop in a given electrolyte against a standard reference electrode. The relative position of two metals on such a series gives a good indication of which metal is more likely to corrode more quickly. However, other factors such as water aeration and flow rate can influence the rate of the process markedly.

The galvanic series orders metal and alloy reactivity in seawater (Fig. 3.1). It is evident that when zinc and iron are coupled, zinc undergoes corrosion because it is more active than iron. Interestingly, however, when iron and copper are coupled, iron undergoes corrosion because it is more active than copper (Fig. 3.2).

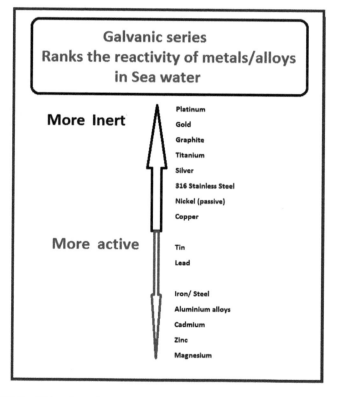

Figure 3.1 Galvanic series.

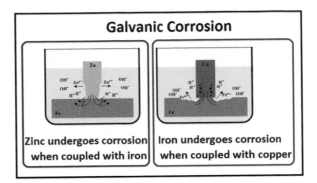

Figure 3.2   Galvanic coupling of zinc and iron, and copper and iron.

3.2   Examples of Galvanic Corrosion

Figure 3.3 shows a stainless-steel cable ladder with mild steel bolts. It can be seen that the mild steel bolts have undergone corrosion.

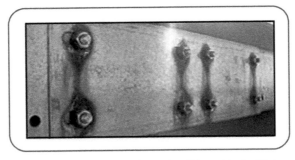

Figure 3.3   Stainless-steel cable ladder with mild steel bolts [1]. Image: D3j4vu at English Wikipedia (https://creativecommons.org/licenses/by-sa/3.0/deed. en).

A sheet of iron or steel covered with a zinc coating is called corrugated iron. Since zinc is less noble, it undergoes corrosion in preference to steel. Interestingly, in the case of a tin can, when the tin coating is broken, steel undergoes corrosion and not tin, since tin is relatively more noble than steel (Fig. 3.4) [1].

Figure 3.4 Rusted corrugated steel roof. Image: https://en.wikipedia.org/ wiki/Corrugated_galvanised_iron (https://en.wikipedia.org/wiki/Corrugated_ galvanised_iron).

3.2.1 Statue of Liberty

Regular maintenance checks discovered that the Statue of Liberty suffered from galvanic corrosion between the outer copper skin and the support structure made of wrought iron (Figs. 3.5 and 3.6).

Figure 3.5 The Statue of Liberty [1]. Image license: https://creativecommons. org/licenses/by-sa/2.0/deed.en.

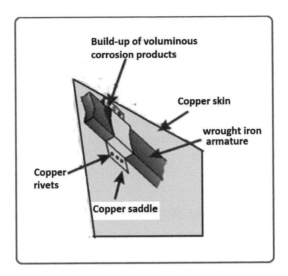

Figure 3.6 Galvanic corrosion in the Statue of Liberty [1].

When the Statue of Liberty was built in the 1880s, the problem of corrosion was predicted. There was an insulation layer of shellac between the two metals. But in due course, the shellac layer failed and the two metals came into contact each other. This resulted in the corrosion of the iron supports. During renovation of the statue, polytetrafluoroethylene (PTFE) was used as an insulator, which has increased the lifetime of the statue [1].

In July 1981, the National Park Service and an international team of engineers and architects conducted a number of inspections of the Statue of Liberty and revealed the following issues that required attention [2]:

- The outer copper skin of the statue had rust stains.
- The paint on the interior copper skin and support structure was peeling.
- The copper skin was highly corroded.
- The torch was degrading.
- The shoulder of the torch arm had structural issues.
- The crown area and spikes were degrading.
- The iron armature was corroding.

The most severe problem that required immediate attention was galvanic corrosion between the iron and copper of the iron armature that was in contact with the copper skin. The electrochemical potential difference between iron and copper was the driving force for electrolysis [2].

The iron armature was completely interconnected, and a single physical contact between the armature and the copper skin was all that was needed for all the copper in the statue to get electrical continuity. The designers had, of course, insulated the two materials with shellac-soaked asbestos cloths. However, this was a temporary barrier, which became porous with time and by the wicking and capillary action that the electrolytic continuity provided [2].

3.2.2 The Royal Navy and the HMS *Alarm*

In England, during the 17th century, it was observed that when lead came in contact with rudder iron, the iron deteriorated. However, many people did not know the reason as at that time the concept of galvanic corrosion was not popular. Admiralty Secretary Samuel Pepys was not an exception. But he was for the removal of lead sheathing from British Royal Navy vessels to prevent the "disintegration of their rudder irons and bolt heads" [1].

According the principle of galvanic corrosion, when copper and iron are in contact with each other in the presence of an electrolyte, iron undergoes corrosion and copper does not. This was observed in the Royal Navy in 1761. Marine vessels were sheathed in copper to reduce marine weed accumulation and protect against shipworm. The Royal Navy fitted the hull of the frigate His Majesty's Service (HMS) *Alarm* with copper plating (Fig. 3.7) [1]. When the ship returned after her voyage to the West Indies, it was observed that the copper part was in good condition, whereas the iron component had undergone corrosion [1].

Interestingly, it was observed that when the copper part and iron nails were separated by water-resistant brown paper, both copper and iron remained in good condition because there was no contact between copper and iron. The brown paper acted as an insulator, which prevented galvanic corrosion. Hence it was reported to the concerned authorities that iron and copper should not come in direct contact in seawater in order to avoid galvanic corrosion [1].

Figure 3.7 HMS *Alarm*, a 32-gun fifth rate Niger-class frigate of the Royal Navy. Image: https://commons.wikimedia.org/wiki/File:HMS_Alarm_(1758).jpg.

3.2.3 US Navy Littoral Combat Ship *Independence*

When aluminum and steel are in contact in the presence of an electrolyte like seawater, aluminum acts as the anode and it corrodes and protects steel. Hence aluminum is often referred to as a "sacrificial anode." This was observed in the latest US Navy attack littoral combat vessel the USS *Independence*. When steel water jet propulsion systems were attached to an aluminum hull, aluminum acted as the anode and underwent corrosion. Had these materials been insulated, aluminum would have been saved without any material loss (Fig. 3.8) [1].

Figure 3.8 US Navy Littoral Combat Ship *Independence*. Image: https://en.wikipedia.org/wiki/USS_Independence_(LCS-2) (https://creativecommons.org/licenses/by-sa/3.0/).

3.2.4 Corroding Light Fixtures

A Big Dig vehicular tunnel (Thomas P. "Tip" O'Neill Jr. Tunnel) is located in Boston. One day a heavy light fixture from the ceiling fell down. Analysis revealed that this was due to galvanic corrosion. Aluminum was in close contact with stainless steel in the presence of salt water (Fig. 3.9) [4].

Figure 3.9 Fall of a heavy light fixture from the ceiling of the Big Dig vehicular tunnel in Boston. Image: http://archive.boston.com/news/local/massachusetts/gallery/ntsb_tunnel_collapse?pg=5.

In 2001, it was well known to the Turnpike Authority officials and contractors that there were many leaks in the ceiling and wall fissures, leading to extensive water damage to the steel supports and fireproofing systems and overloaded drainage systems. As many as 400 leaks riddled the tunnel system. Due to a major leak, on September 15, 2004, the Interstate 93 north tunnel was closed for conducting repairs.

The presence of highly corrosive salt water was noticed in the leakage. This water came from Boston Harbor and the Atlantic Ocean. The most corrosive chloride ions present in the brackish water were responsible for corroding the ceiling light fixtures. The rebars and other structural steel reinforcements holding the tunnel walls and ceiling in place corroded [3]. This resulted in the collapse of the ceiling. The formation of rust on the steel rebars in concrete resulted in an increase of volume and hence falling of the ceiling as a natural consequence. The inspecting safety team identified as many as 242 potentially dangerous bolt fixtures supporting the ceiling tiles, and they were corrected [4].

3.2.5 A Lasagna Cell

A lasagna is a type of wide, flat pasta. It is a salty, moist food item. It can be stored in a steel baking pan and covered with aluminum foil. When lasagna is covered with aluminum foil, a "lasagna cell" is created. Small holes are observed on the aluminum foil because of galvanic corrosion. Corroded aluminum can be seen on the food. In this example, the aluminum foil acts as the anode and undergoes corrosion. The steel backing acts as the cathode. It does not undergo corrosion. The salty, moist food acts as the electrolyte [1]. It is always better if the backing material is also made of aluminum. An aluminum–aluminum arrangement is always better and prevents galvanic corrosion [1].

3.2.6 Electrolytic Cleaning

In electrolytic cleaning, silverware is cleaned. The silver and a piece of aluminum are immersed in a sodium bicarbonate solution, which functions as an electrolyte. Silver oxide is stripped off. This is an example of galvanic corrosion [1]. Silver-coated base metals should be avoided because the base metal (stainless steel, for example) will undergo galvanic corrosion.

3.3 Preventing Galvanic Corrosion

Galvanic corrosion can be prevented by many methods [1]. They are discussed in this section.

3.3.1 Insulation of Two Metals from Each Other

In galvanic corrosion, the electrical contact plays an important role. If the contact is eliminated, then galvanic corrosion will not take place. This can be achieved by introducing an insulating material between the two dissimilar metals with different electrode potentials. A spool of pipe made of plastic materials can be used to isolate piping. The spool should be of a sufficient length to be effective [1].

3.3.2 Hull Connected to Earth

For safety reasons, metal boats connected to a shore line electrical power feed have to have the hull connected to earth. The use of a galvanic isolator is recommended. This consists of two semiconductor diodes in series, in parallel with two diodes conducting in opposite directions in order to prevent current flow while the applied voltage is less than 1.4 V but to allow full current flow in the case of an electrical fault. However, minor leakage of current will still occur through the diodes, resulting in slightly faster corrosion than normal [1].

3.3.3 No Contact with an Electrolyte: Using Hydrophobic Compounds

Contact with an electrolyte can be avoided by using water-repellent hydrophobic compounds, for example, grease or a protective layer of coating such as paint or varnish. Care should be taken to increase the anodic area and decrease the cathodic area. If the anodic area is very small, it will lead to severe corrosion [1].

3.3.4 Use of Antioxidant Paste

Antioxidant paste can be used to prevent galvanic corrosion, for example, between copper and aluminum. Usually, the paste is made up of a lower-nobility metal than copper or aluminum [1].

3.3.5 Use of Metals That Have Similar Electropotentials

Galvanic corrosion can be prevented by making use of metals that have more or less the same electropotentials. In such cases, the galvanic current will be less. Thus, galvanic corrosion may be prevented. The best way is to use the same metal instead of dissimilar metals with larger potential differences [1].

3.3.6 Electroplating

Steel-based materials can be protected by electrogalvanization. Zinc functions as the sacrificial anode. More noble metals, such as chromium and nickel, which resist corrosion, can be used. Zinc or aluminum can be used as sacrificial anodes to protect structures made of steel (Figs. 5.1 and 5.2, Chapter 5) [1].

3.3.7   Cathodic Protection

To protect buried metallic structures, cathodic protection (CP) is used. The metal structure to be protected is the cathode. Metals such as zinc, magnesium, and aluminum can be used as the anode. The anode undergoes corrosion, whereas the cathode is protected, hence the term "cathodic protection." In this process, electrons from the anode are pumped on the cathode, which prevents the loss of electrons from the cathode, and thus corrosion is prevented. CP finds application in preventing corrosion in water heaters, hulls of ships, and buried structures [1, 5]. For more details on CP, see Chapter 4.

3.3.8   Proper Designing

Galvanic corrosion between dissimilar metals and/or alloys, for example, steel and aluminum, can be prevented by proper insulation between the two metals (Fig. 3.10).

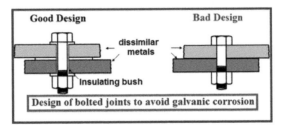

Figure 3.10   Proper designing to avoid galvanic corrosion.

3.4   The Anodic Index

The anodic index is a parameter that is a measure of the electrochemical voltage that develops between a metal and gold. The relative voltage of a pair of metals is determined by subtracting their anodic indices. The anodic index helps in predicting the compatibility of two different metals.

The difference in the anodic index between the two metals should be less than 0.25 V for normal environments. It should not be more than 0.15 V for harsh environments, such as outdoors, high-humidity, and salt environments. For environments in which temperature and humidity are controlled, an anodic index of 0.50 V can be tolerated. The base materials can be protected from corrosion by finishing and plating, which allow the dissimilar materials to be in contact but protect the base materials from corrosion. In this case, the metal with the most negative anodic index will undergo corrosion. "This is why sterling silver and stainless-steel tableware should never be placed together in a dishwasher at the same time, as the steel items will likely experience corrosion by the end of the cycle (soap and water having served as the chemical electrolyte and heat having accelerated the process)" [1] (Table 3.1).

Table 3.1 Compatibility of two different metals predicted by the anodic index [1]

METAL	INDEX (V)
Most cathodic	
Gold, solid and plated, gold-platinum alloy	−0.00
Rhodium plated on silver-plated copper	−0.05
Silver, solid or plated; monel metal; high-nickel-copper alloys	−0.15
Nickel, solid or plated; titanium an its alloys; monel	−0.30
Copper, solid or plated; low brasses or bronzes; silver solder; German silvery high-copper-nickel alloys; nickel-chromium alloys	−0.35
Brasses and bronzes	−0.40
High brasses and bronzes	−0.45
18% chromium-type corrosion-resistant steels	−0.50
Chromium plated; tin plated; 12% chromium-type corrosion-resistant steels	−0.60
Tin plated; tin-lead solder	−0.65
Lead, solid or plated; high-lead alloys	−0.70
2000 series wrought aluminum	−0.75
Iron, wrought, gray or malleable, plain carbon and low-alloy steels	−0.85

METAL	INDEX (V)
Aluminum, wrought alloys other than 2000 series aluminum, cast alloys of the silicon type	−0.90
Aluminum, cast alloys other than silicon type, cadmium, plated and chromate	−0.95
Hot-dip-zinc plate; galvanized steel	−1.20
Zinc, wrought; zinc-base die-casting alloys; zinc plated	−1.25
Magnesium and magnesium-base alloys, cast or wrought	−1.75
Beryllium	−1.85
Most anodic	

References

1. https://en.wikipedia.org/wiki/Galvanic_corrosion (https://en.wikipedia.org/wiki/Wikipedia:Text_of_Creative_Commons_Attribution-ShareAlike_3.0_Unported_License)

2. https://corrosion-doctors.org/Landmarks/Statue.htm

3. https://en.wikipedia.org/wiki/Big_Dig (https://en.wikipedia.org/wiki/Wikipedia:Text_of_Creative_Commons_Attribution-ShareAlike_3.0_Unported_License)

4. https://en.wikipedia.org/wiki/Big_Dig_ceiling_collapse (https://en.wikipedia.org/wiki/Wikipedia:Text_of_Creative_Commons_Attribution-ShareAlike_3.0_Unported_License)

5. https://en.wikipedia.org/wiki/Cathodic_protection (https://en.wikipedia.org/wiki/Wikipedia:Text_of_Creative_Commons_Attribution-ShareAlike_3.0_Unported_License)

Chapter 4

Industrial Methods of Preventing Galvanic Corrosion: Part I

Galvanic corrosion, or bimetallic corrosion, can be prevented by many industrial processes, such as galvanization, cathodic protection, and anodic protection.

This chapter discusses galvanization in detail.

4.1   Introduction

The rusting of steel and iron can be controlled by coating them with zinc. This process is called galvanization. This process mainly prevents metals from corrosion. Galvanization can be done by dipping the material to be protected in a molten zinc bath [1]. This process is called hot-dip galvanization [1]. Hot-dip galvanization will be explained in detail later.

Galvanization protects the base metal in three ways:

- It forms a coating of zinc that, when intact, prevents corrosive substances from reaching the base steel or iron.
- The zinc serves as a "sacrificial" anode (it corrodes first) so that even if the coating is scratched, the exposed steel will still be protected by the remaining zinc.
- Sometimes, chromates are applied over the zinc.

Titanic Corrosion
Susai Rajendran and Gurmeet Singh
Copyright © 2020 Jenny Stanford Publishing Pte. Ltd.
ISBN 978-981-4800-95-2 (Paperback), 978-1-003-00449-3 (eBook)
www.jennystanford.com

4.2   History and Etymology of Galvanization

"The earliest-known example of galvanized iron was encountered by Europeans on 17th-century Indian armor in the Royal Armouries Museum collection" [1]. The term was named in English via French after the Italian scientist Luigi Galvani (Fig. 4.1).

Figure 4.1   Luigi Galvani. Image: https://commons.wikimedia.org/w/index.php?curid=2707375.

In the 19th century, the term "galvanizing" was used to describe the administration of electric shocks; this was also called *Faradism*. This usage is the origin of the metaphorical use of the verb "galvanize," such as to "galvanize into action," meaning stimulating a complacent person or group to take action.

In 1837, Stanislas Sorel of Paris patented a precursor to hot-dip galvanization, which is known as galvanic paint. In modern days, the term "galvanizing" refers to a coating of zinc [1].

4.3   Methods of Galvanization

4.3.1   Hot-Dip Galvanization

Hot-dip galvanization is the process of coating iron and steel with a layer of zinc by immersing the metal in a bath of molten zinc at a temperature of around 449°C (840°F).

During the process of hot-dip galvanization, a metallurgical bond is formed between zinc and steel. The use of materials before and after galvanization remains the same. However, the corrosion resistance of the materials increases after coating [2]. Hot-dip galvanization includes the following steps (Fig. 4.2):

1. Cleaning/decreasing: Steel is treated with concentrated sodium hydroxide to remove paint, oil, and grease [2].
2. Rinsing: The material is washed with a minimum amount of pure water [2].

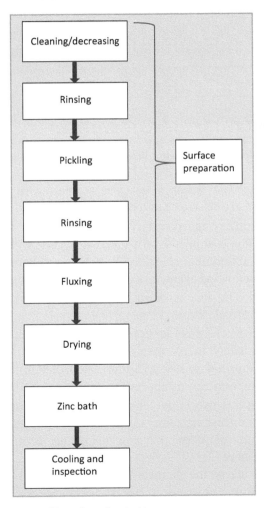

Figure 4.2 Process of hot-dip galvanization.

3. Pickling: When the material is treated with concentrated HCl, scales are removed (Fig. 4.3) [2]. This is called "pickling." During the pickling process, rust and scales are removed (https://en.wikipedia.org/wiki/Pickling_(metal)).

Figure 4.3 Mill scale on an anvil. Image: https://en.wikipedia.org/wiki/ Galvanization (http://www.gnu.org/licenses/old-licenses/fdl-1.2.html).

4. Rinsing: The material is again washed with a minimum amount of pure water to remove the pickling solution [2].
5. Fluxing: The cleaned surface may be oxidized by air. To prevent this, a flux is applied on the metal surface. A well-known flux is zinc ammonium chloride [2]. A flux is a chemical cleaning agent, a flowing agent, or a purifying agent.
6. Drying: The flux is dried on the steel. This helps in the deposition of zinc on the metal [2].
7. Zinc bath: The steel is dipped into a molten zinc bath. Lead is often added to a molten zinc bath to control the amount of zinc deposited on steel (Fig. 4.4). Environmental regulations in many countries disapprove the use of lead [2].
8. Cooling and inspection: Finally, the steel is cooled in a quench tank, reducing its temperature and preventing undesirable reactions between the newly formed coating and the atmosphere [2].

Figure 4.5 shows the characteristic spangle (grains, a visible and aesthetic feature, in galvanized coatings) of zinc–iron alloy formed by hot-dip galvanization.

Figure 4.4 Aluminum dross. Image: https://en.wikipedia.org/wiki/Dross.

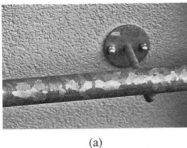

(a)   (b)

Figure 4.5 Characteristic spangle of hot-dip galvanization on (a) a steel hand rail (https://creativecommons.org/licenses/by-sa/3.0/de/deed.en) and (b) a street lamp in Singapore (https://creativecommons.org/licenses/by/2.5/).

A hot-dip-galvanized steel strip is often referred to as galvanized iron (GI). It is mainly used for applications that need the strength of steel, combined with the corrosion resistance of zinc. Common examples are handrails, roofing, safety barriers, and automotive body parts (Fig. 4.6) [2].

Figure 4.6 Corrugated galvanized iron roofing in Mount Lawley, Western Australia. Image: https://en.wikipedia.org/wiki/Corrugated_galvanised_iron (https://creativecommons.org/licenses/by/2.5/).

Batch galvanizing is hot-dip galvanization in which individual metal articles are galvanized. For example, wrought iron gates can be galvanized by batch galvanizing. Recently, the electrogalavanizing process has been introduced. This is considered an electroplating process. The film produced in this process is much thinner but stronger [2].

4.3.2 Electroplating

In the electrogalvanizing process, a thin layer of zinc is coated onto steel or any metal. This coated film protects the coated material from corrosion. In electroplating, zinc is used as the anode, while the object to be electroplated is used as the cathode. The electrolyte is a zinc salt solution (Fig. 4.7) [3].

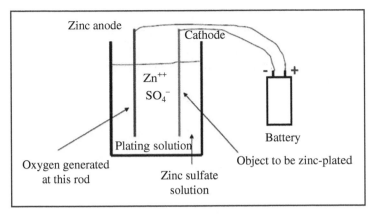

Figure 4.7 Principle of zinc plating.

In hot-dip galvanization, the metal to be coated is immersed in molten zinc. As it is withdrawn, the zinc cools and forms a coating on the metal. In contrast, in electroplating, the metal is immersed in an aqueous bath and an electric current is passed through the aqueous solution, which moves electrons from the anode to the cathode. This oxidizes the zinc (anode), which dissolves as zinc ions in the aqueous solution. The zinc ions then move to the metal (anode) and are reduced on it in the form or metallic zinc. Hot-dip-galvanized coatings are almost always many times thicker.

It is strongly felt that electroplating of zinc is more advantageous than hot-dip galvanization. Some of the advantages are listed below:

- The film is thin but stronger.
- A color coating is possible.
- The coating is found to be brighter.
- It has aesthetical value.

4.3.2.1 Hot-dip galvanization vs. electroplating

For a steel surface, hot-dip galvanization is selected. In this case, the zinc film coated on steel is thick. Usually, automobile bodies are given decorative coatings. In these cases, first zinc is electroplated and then decorative coatings are created by electroplating.

The products made by hot-dip galvanization are not sufficiently strong. These materials undergo corrosion in the presence of strong acids and also acid rain. It is found that electrogalvanization is better than hot-dip galvanization. Therefore, for bolts and nuts, electrogalvanization is preferred to hot-dip galvanization. This is also the case with cars and bicycles [1].

Electroplated zinc coatings are smooth and shiny and preferable for aesthetic reasons, whereas galvanized coatings may be spangled (gray and drippy) (Fig. 4.8). Zinc electroplating is thin and usually does not cause any problems with fasteners (except extremely small ones). On the other hand, galvanized coatings are heavy and interfere with fastener threads unless specially dimensioned to consider the coating thickness.

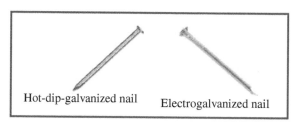

Hot-dip-galvanized nail   Electrogalvanized nail

Figure 4.8   Hop-dipped-galvanized versus electrogalvanized nails.

The finish of electrogalvanization is readily paintable, which is an important quality to manufacturers of trailers, appliances, agriculture equipment, etc.

4.3.3   Sherardizing

The process of galvanization of ferrous metal surfaces is called "sherardizing." It involves thermal diffusion galvanizing. Sherardizing is called by different names such as vapor galvanizing and dry galvanizing. The process was invented by British metallurgist Sherard Osborn Cowper-Coles in 1900.

Steel parts, zinc dust, and sand are taken in a rotating drum. The mixture is heated strongly. At a temperature above 300°C, zinc evaporates and diffuses into the steel substrate. Diffusion-bonded Zn-Fe phases are produced.

4.3.3.1   Advantages of sherardizing

- Sherardizing is suitable for small parts and parts whose inner surfaces need to be coated.

- Batches of small items can be sherardized. Part size is limited by the drum size.
- Pipes up to 6 m in diameter needed for the oil industry can be sherardized.
- No pretreatment is needed if the metal surface is free from oxides or scales.
- The process is hydrogen free, so hydrogen embrittlement does not occur (Fig. 4.9) [4].
- No liquids are involved, so this process is also known as "dry galvanizing." It can avoid possible problems caused by hydrogen embrittlement.
- The dull-gray crystal structure of the zinc diffusion coating has good adhesion to paint, powder coatings, and rubber.
- It is a preferred method for coating small, complex-shaped metals
- Sherardizing is suitable for smoothing rough surfaces on sintered metal items [1].

Figure 4.9 Sherardized objects. Image: https://www.pinterest.com/pin/487796203367752825/.

There are two standards for the process of sherardizing:

- BS EN 13811: *2003 Sherardizing. Zinc diffusion coatings on ferrous products. Specification.*

- BS EN ISO 14713-3: *2009 Zinc coatings. Guidelines and recommendations for the protection against corrosion of iron and steel in structures. Part 3. Sherardizing.*

4.3.3.2 Hot-dip galvanization vs. sherardizing

In both hot-dip galvanization and sherardizing, zinc is ultimately coated on a metal surface. Sherardizing is a difficult method, but hot-dip galvanization is easy. The two differ mainly in the method of application. The galvanic couple between hot-dip-galvanized zinc and steel is far stronger than in sherardized zinc. This is due to the fact during hot-dip galvanization, there is strong evidence to prove the formation of an iron–zinc alloy (https://www.finishing.com/74/60.shtml).

4.4   Uses of Galvanization

4.4.1   Galvanized Construction Steel

Construction steel is the most common use for galvanized metal, and annually, hundreds of thousands of tons of steel products are galvanized around the world. In developed countries, there are many galvanizing factories, and many steel items are galvanized for protection, for example, street furniture, building frameworks, balconies, staircases, and walkways [1].

4.4.2   Galvanized Piping

Before the 20th century, cast iron and lead were used in cold water plumbing. Galvanized piping was introduced in the early part of the 20th century. Rusting begins from the inside of the pipe, forming layers of plaque. Finally, pipe failure occurs because of the formation of sludge and scales. Depending on the water parameters and soil quality, the lifetime of GI pipes is about 70 years. The lifetime also depends on the thickness of zinc coatings.

After the Second World War, copper and plastic pipes began to be used in the interior drinking water service, although galvanized steel pipes are still used for outdoor applications.

GI pipes are preferred for houses. The lifetime of GI pipes can be increased by an epoxy resin coating [1].

4.5 Eventual Corrosion

Galvanization is not a permanent solution to prevent corrosion due to environmental attack. Rusting is inevitable after a few decades of exposure to weather. The exposure is rather severe in an acidic environment. Corrugated iron sheet roofing starts to degrade within a few years, even after a zinc coating. This is quite severe in marine and salty environments, where the high electrical conductivity of seawater increases the rate of corrosion. Due to road salt, galvanized car frames corrode much faster in cold environments.

Galvanizarion of steel will be an everlasting solution, provided it is supported by paint coatings and additional sacrificial anodes. In an environmentally friendly atmosphere, free from toxic gases such as carbon dioxide and sulfur dioxide, galvanized steel can withstand corrosion for more than 100 years [1].

References

1. https://en.wikipedia.org/wiki/Galvanization (https://en.wikipedia.org/wiki/Wikipedia:Text_of_Creative_Commons_Attribution-ShareAlike_3.0_Unported_License)

2. https://en.wikipedia.org/wiki/Hot-dip_galvanization (https://en.wikipedia.org/wiki/Wikipedia:Text_of_Creative_Commons_Attribution-ShareAlike_3.0_Unported_License)

3. https://en.wikipedia.org/wiki/Electrogalvanization (https://en.wikipedia.org/wiki/Wikipedia:Text_of_Creative_Commons_Attribution-ShareAlike_3.0_Unported_License

4. https://en.wikipedia.org/wiki/Sherardising (https://en.wikipedia.org/wiki/Wikipedia:Text_of_Creative_Commons_Attribution-ShareAlike_3.0_Unported_License)

Chapter 5

Industrial Methods of Preventing Galvanic Corrosion: Part II

This chapter discusses two more industrial methods of preventing galvanic corrosion, cathodic protection and anodic protection.

5.1 Cathodic Protection

Cathodic protection (CP) is defined as a method of controlling corrosion of a metal by making it the cathode in an electrochemical cell. By CP, metals and metal structures buried in the ground and ship hulls can be protected from corrosion. In this method, electrons are pumped on the metal surface to be protected from another metal that readily releases electrons. The former acts as the cathode, while the latter acts as the anode. That is, the material to be protected is forced to act as the cathode. The metal that releases electrons will undergo corrosion and is called the "sacrificial anode" (Figs. 5.1 and 5.2) [1]. In some cases, sufficient current is supplied by an external direct current (DC) electrical power source. This method is called impressed current CP. Long pipelines can be protected by this method [1].

CP systems are used to protect various metallic structures in different environments, for example, steel water or fuel pipelines and steel storage tanks (i.e., home water heaters), ship and boat hulls, offshore oil platforms, and metal reinforcement bars in concrete

Titanic Corrosion
Susai Rajendran and Gurmeet Singh
Copyright © 2020 Jenny Stanford Publishing Pte. Ltd.
ISBN 978-981-4800-95-2 (Paperback), 978-1-003-00449-3 (eBook)
www.jennystanford.com

buildings and structures. In galvanized steel, zinc protects steel from corrosion by the sacrificial anode principle. Stress corrosion cracking can also be prevented by CP [1].

Figure 5.1 Sacrificial anodes (light-colored rectangular bars) made of aluminum, mounted on a steel jacket structure [1] (https://creativecommons.org/licenses/by-sa/2.5/).

Figure 5.2 Sacrificial anode (rounded object) made of zinc, screwed to the underside of a small boat's hull [1] (https://creativecommons.org/licenses/by-sa/3.0/).

5.1.1 Types of Cathodic Protection

5.1.1.1 Passive cathodic protection

In this method, a more active metal is attached to the metal surface to be protected. Both metals are in contact with each other in the presence of an electrolyte (Fig. 5.3) [1]. The potential of the steel surface is polarized to be the negative side until the surface has a uniform potential. Now the more active metal undergoes corrosion, and the metal structure is protected from corrosion. Finally the more active metal has to be replaced by a new one.

Electrons flow from the anode to the cathode, which is the material to be protected. For this electron flow to be effective, a good electrically conductive contact is essential. Examples of sacrificial anodes are alloys of zinc, magnesium, and aluminum. The composition of sacrificial anodes has been standardized by a series of experiments.

For effective CP, the anode should have a low electrode potential and the cathode (the metal to be protected) should have a high electrode potential. A galvanic series that is used to select the anode metal is shown in Table 5.1. In the series, the anode lies below the metal to be protected [1].

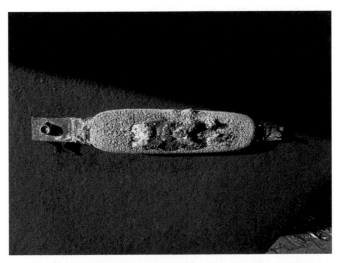

Figure 5.3 Galvanic sacrificial anode attached to the hull of a ship; the anode is showing corrosion [1] (https://creativecommons.org/licenses/by-sa/3.0/).

Table 5.1 Simplified galvanic series is used to select the anode metal [1]

Metal	Potential with respect to a Cu:CuSO$_4$ reference electrode in neutral pH environment (volts)
Carbon, graphite, coke	+0.3
Platinum	0 to −0.1
Mill scale on steel	−0.2
High-silicon cast iron	−0.2
Copper, brass, bronze	−0.2
Mild steel in concrete	−0.2
Lead	−0.5
Cast iron (not graphitized)	−0.5
Mild steel (rusted)	−0.2 to −0.5
Mild steel (clean)	−0.5 to −0.8
Commercially pure aluminum	−0.8
Aluminum alloy (5% zinc)	−1.05
Zinc	−1.1
Magnesium alloy (6% Al, 3% Zn, 0.15% Mn)	−1.6
Commercially pure magnesium	−1.75

5.1.1.2 Impressed current systems

In some cases, the structure to be protected may be very large, and in some areas, the electrolyte resistivity may be high. That is, the flow of electrons from the anode to the cathode will not be easy. In such cases, another method is used. It is called the impressed current cathodic protection (ICCP) system (Fig. 5.4). In this method, the anodes are connected to a DC power source, usually a transformer-rectifier connected to alternating current (AC) power. Alternative power sources such as wind power, solar panels, or gas-powered thermoelectric generators may be used [1].

High-silicon cast iron, graphite, mixed-metal oxides, platinum- and niobium-coated wires, and other materials are used as anodes in ICCP systems. The anodes are available in different shapes such as tubular or solid rods.

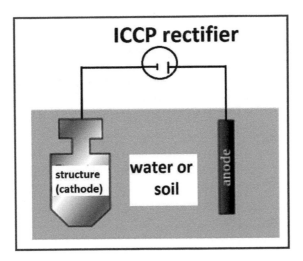

Figure 5.4 Impressed current cathode protection system. A DC source drives the protective electrochemical reaction [1] (https://creativecommons.org/licenses/by-sa/3.0/).

Depending on many design and field condition factors, anodes are arranged in ground beds, either distributed or in a deep vertical hole, to protect corrosion of pipelines. The custom-manufactured CP transformer-rectifier units are provided with a wide variety of features, such as remote monitoring and control, integral current interrupters, and various types of electrical enclosures. The structure to be protected is connected to the output DC negative terminal.

The rectifier output DC positive cable is connected to the anodes. The AC power cable is connected to the rectifier input terminals. The voltage output of the ICCP system can be selected with taps on the transformer windings and jumper terminals designed on the CP transformer-rectifier units.

CP transformer-rectifier units used in water tanks and also used in other applications are made with solid-state circuits. This enables them to automatically adjust the operating voltage to maintain the optimum current output or structure–electrolyte potential.

To show the operating voltage (DC and sometimes AC) and current output, analog or digital meters are often installed.

For large, complex target structures and seashore structures, ICCP systems are often designed with multiple independent zones of anodes with separate CP transformer-rectifier circuits [1].

5.1.2 Applications

In 1824, HMS *Samarang* used CP to protect the ship from marine corrosion. Sacrificial anodes made of iron were attached to the copper sheath of the hull below the waterline. This reduced the corrosion of copper to a great extent. But there was an increase in marine growth. That is, when the corrosion of copper was prevented, the amount of copper ions released reduced. Usually, copper ions have an antifouling effect. Lack or copper ions decreased the antifouling effect. Excess of marine growth decreased the efficiency of the ship. The question before the Royal Navy was whether to allow copper corrosion and prevent marine growth or to prevent copper corrosion at the cost of marine growth. Finally, the Royal Navy opted for copper corrosion and dropped the idea of CP [1].

5.1.2.1 Pipelines

Pipelines carrying dangerous products are protected by a protective coating, followed by CP. In the ICCP system, the pipeline is attached to a DC power source, usually an AC-powered transformer-rectifier and anodes. The anodes are buried in the ground (Fig. 5.5) [1]. The site is usually referred to as the "anode ground bed."

The positive DC output terminal would be connected through cables to the anode array. Another cable would connect the negative terminal of the rectifier to the pipeline. This is usually done through junction boxes. Depending on the size of the pipeline and coating quality, generally, the DC power source would typically have a DC output of up to 50 A and 50 V.

The sacrificial anodes could be buried in the ground bed consisting of a vertical hole, backfilled with conductive coke. This backfill will improve the lifetime and performance of the anodes. Sometimes, the anodes are placed in a prepared trench, surrounded by conductive coke and backfilled. Depending on the application, location, and soil resistivity, the choice of ground bed type and size varies.

After performing various tests, such as measurements of pipe-to-soil potentials or electrode potential, the DC CP current is then adjusted to the optimum level. When the pipelines are of a small diameter and of limited length, the pipelines are protected by sacrificial anodes. This method is more economical. The CP current is sent from the anode to the structure being protected.

When the owners feel that the cost is reasonable for the expected pipeline service life, water pipelines of various pipe materials are also provided with CP [1].

Figure 5.5 Cathodic protection markers over a gas pipeline in Leeds, West Yorkshire, England [1] (https://creativecommons.org/licenses/by-sa/3.0/).

5.1.2.2 Ships and boats

ICCP is used for large ships. For ordinary ships, galvanic anodes are attached to the hull. In due course, the galvanic anodes can be easily replaced during inspection and maintenance of the ships, during which process, the ships are removed from the water (Fig. 5.6) [1].

Usually, galvanic anodes are shaped so as to reduce drag in the water. They are fitted flush to the hull to minimize drag.

Smaller vessels, such as yachts, have nonmetallic hulls. To protect areas such as outboard motors, they are fitted with galvanic anodes. There should be a solid electrical connection between the anode and the item to be protected.

Figure 5.6 Zinc block sacrificial anodes (white patches visible on the ship's hull) [1]. Source: US government.

Platinized titanium is an inert material. The anodes for ICCP on ships are constructed of platinized titanium. The anodes are mounted on the outside of the hull. A DC power supply is supplied within the ship.

The anode cables are introduced into the ship through a compression seal fitting and routed to the DC power source. The negative cable from the power supply is attached to the hull to complete the circuit. Ship ICCP anodes are flush-mounted. This minimizes the effects of drag on the ship. To avoid mechanical damage, the anodes are located a minimum 5 ft. below the light load line in an area.

In some ships, there are aluminum hulls with steel fixtures. In these cases, the aluminum hull will act as a galvanic anode. So corrosion is enhanced. In such cases, aluminum or zinc galvanic anodes could be used to offset the potential difference between the aluminum hull and the steel fixture. In case the steel fixtures are large, many galvanic anodes may be required for even a small ICCP system [1].

5.1.2.3 Marine cathodic protection

In the marine environment, there are several types of structures to be protected. These structures include jetties, harbors, and offshore structures. Generally, galvanic anodes are used. Sometimes, ICCP can also be used [1].

5.1.2.4   Steel in concrete

The use of CP for concrete reinforcement is slightly different. In concrete reinforcement, the anodes and reference electrodes are generally embedded in the concrete at the time of construction during the pouring of concrete. The usual technique for concrete buildings, bridges, and similar structures is to use ICCP. However, there are systems available that use the principle of galvanic CP as well.

In a typical atmospherically exposed concrete structure such as a bridge, there will be many more anodes distributed through the structure as opposed to an array of anodes as used on a pipeline. In such a case, an automatically controlled DC power source is used. There is remote monitoring also. For buried or submerged structures, the treatment is similar to that of any other buried or submerged structure.

There are pipelines constructed from prestressed concrete cylinder pipe (PCCP). The techniques used for CP are usually as for steel pipelines, except that the applied potential must be limited to prevent damage to the prestressing wire.

The steel wire in a PCCP is stressed. It may undergo corrosion and it may result in failure. There is an additional problem. An excessively negative potential can cause excessive hydrogen ions. This will result in hydrogen embrittlement of the wire, resulting in failure.

The failure of many such wires will result in catastrophic failure of the PCCP. Hence it is inferred that implementation of ICCP requires careful control to ensure satisfactory protection. So it is concluded that a simpler option is to use galvanic anodes, which are self-limiting and need no control [1].

5.1.2.5   Internal cathodic protection

CP can be used to prevent corrosion on the internal surfaces of pipelines, vessels, and tanks that are utilized to store or transport liquids. For this purpose, galvanic systems and ICCP can be used. Water storage tanks and power plant shell and tube heat exchangers are protected from corrosion by internal CP [1].

5.1.2.6 Galvanized steel

In galvanization, zinc is coated on steel. This can be done by hot-dip galvanization or electrogalvanizing. Galvanized materials are stable in most of the environments. They have the synergistic effect of coatings and the benefits of CP. Even if the zinc coating is scratched or locally damaged, steel will be protected from corrosion. This is due to the fact that during this process, zinc will act as the anode and steel will function as the cathode because of galvanic action and thus steel will be protected. Zinc will undergo "sacrificial corrosion."

Galvanization uses the principle of CP. But it is not true CP. In CP, the anode and cathode are connected through an electrolyte. But in the case of galvanization, there is no role of the electrolyte. Zinc and steel are in close contact. Thus the area with which zinc is in contact will be protected. In CP, even if the metal is far away and if the metal is connected by an electrolyte, the metal will be protected from corrosion [1].

5.1.2.7 Automobiles

Many companies market electronic devices claiming to mitigate corrosion for automobiles and trucks. But corrosion scientists object to this concept because this claim is not experimentally proved.

In 1996, David McCready sold devices that he claimed protected cars from corrosion. But the Federal Trade Commission (FTC) ordered him to pay restitution. Further, the commission banned the names "Rust Buster" and "Rust Evader" [1].

5.2 Anodic Protection

Whereas in CP, the material to be protected is forced to act as the cathode, in anodic protection (AP), the material to be protected is forced to act as the anode of an electrochemical cell. The electrode potential is controlled in a zone where the metal is passive.

AP is used to protect metals that exhibit passivation in environments. The current density in the freely corroding state could be significantly higher than the current density in the passive state over a wide range of potentials.

AP is possible even in extreme pH environments. Carbon steel tanks containing concentrated sulfuric acid and also 50% sodium hydroxide can be protected from corrosion by AP. For such cases CP is not advisable because of high-current requirements.

In an AP system, an external power supply is connected to auxiliary cathodes and controlled by a feedback signal from one or more reference electrodes.

For AP systems, careful design and control are required. There are many reasons for this. For example, when passivation is lost or unstable, there will be excessive current. This will accelerate corrosion.

5.3 Differences between Cathodic Protection and Anodic Protection

Table 5.2 lists the differences between CP and AP.

Table 5.2 Cathodic versus anodic protection

Cathodic protection	Anodic protection
Applicable to all metals	Applicable to metals that show active-passive behavior
Used where there is no source of power, by employing sacrificial anodes	Can handle more aggressive corrodents
Low installation costs	Low operating costs but high installation costs
Standard and well-established method	Feasibility predicted in the laboratory and easy design

References

1. https://en.wikipedia.org/wiki/Cathodic_protection   (https://en.wikipedia.org/wiki/Wikipedia:Text_of_Creative_Commons_Attribution-ShareAlike_3.0_Unported_License)

2. https://en.wikipedia.org/wiki/Anodic_protection   (https://en.wikipedia.org/wiki/Wikipedia:Text_of_Creative_Commons_Attribution-ShareAlike_3.0_Unported_License)

Chapter 6

Metallurgical Analysis of RMS *Titanic*

6.1 Did a Metallurgical Failure Cause a Night to Remember?

The Royal Mail Ship (RMS) *Titanic* began her maiden voyage on April 10, 1912, from Southampton (England) to New York (USA). Two days later, on April 12, 1912, at 11:40 p.m. Greenland time, the *Titanic* hit an iceberg because of which six of her forward compartments flooded (Table 6.1), sinking her 2 h 40 min later and leading to an estimated 1500 deaths. An often mentioned reason for this disaster is the quality of steel used in the *Titanic*'s construction [1, 2].

Table 6.1 Compartments damaged in the *Titanic*'s collision with the iceberg [1]

Compartment	Area damaged
Fore peak	$0.056\ m^2$
Cargo hold 1	$0.139\ m^2$
Cargo hold 2	$0.288\ m^2$
Cargo hold 3	$0.307\ m^2$
Boiler room 5	$0.121\ m^2$
Boiler room 6	$0.260\ m^2$
Total area damaged	**$1.171\ m^2$**

Titanic Corrosion
Susai Rajendran and Gurmeet Singh
Copyright © 2020 Jenny Stanford Publishing Pte. Ltd.
ISBN 978-981-4800-95-2 (Paperback), 978-1-003-00449-3 (eBook)
www.jennystanford.com

Metallurgical analysis of the steel taken from the hull of the *Titanic*'s wreckage was performed, and investigators found that the steel had a high ductile-to-brittle transition temperature, which made it unsuitable for use in the cold. At the time the *Titanic* collided with the iceberg, the seawater temperature was −2°C. However, analysis also showed that the steel used in the ship was the best plain carbon ship plate available at the time [1].

This chapter discusses the metallurgical analysis that was performed on the remains of the *Titanic*.

6.1.1   The Ship Broke into Two Parts

The survivors of the disaster disagreed as to whether the *Titanic* broke into two parts as it sank, whether she broke into three parts, or whether she sank intact. The ship was found on the ocean floor at a depth of 3700 m by Robert Ballard on September 1, 1985 [3]. Ballard reported that the ship had broken into two major sections, approximately 600 m apart.

6.2   Composition of the Steel

The *Titanic*'s decks were actually steel plates riveted together. Because of this, there were overlaps in the plates and protruding rivet heads, which needed to be covered for leveling them out, making them into a durable floor, and insulating them. Therefore, the steel decks were covered with Litosilo deck sheathing.

In 1996, during an expedition to the wreckage, researchers brought back some hull steel from the *Titanic* for metallurgical analysis. First, the composition of the steel was determined by chemical analysis (Table 6.2). Researchers found a high oxygen content, low silicon content, slightly higher than normal phosphorus content, high sulfur content, and low manganese content. The Mn:S ratio was 6.8:1, which is quite low by modern standards. At low temperature, a high phosphorus, oxygen, and sulfur content can make steel brittle.

Table 6.2   Composition of the steel from the *Titanic* [1]

	C	Mn	P	S	Si	Cu	O	N	Mn:S ratio
Titanic's hull plates	0.21	0.47	0.045	0.069	0.017	0.024	0.013	0.0035	6.8:1

Source: Reprinted by permission from Springer Nature Customer Service Centre GmbH: Springer Nature, *JOM Journal of the Minerals, Metals and Materials Society*, Ref. [1], Copyright (1998).

6.3   Metallography

Optical microscopic examination of the hull plates of the *Titanic* showed banding. Specimens were cut from the hull plate along both transverse and longitudinal directions. Banding was more severe in the longitudinal sections than the transverse sections. There were large masses of MnS particles elongated in the direction of the banding. The average grain diameter was 60.40 μm longitudinally and 41.92 μm transversely. Figure 6.1 shows the microstructure of the *Tiitanic*'s steel.

Figure 6.2 shows a scanning electron microscopy (SEM) micrograph of the polished and etched surface of steel from the *Titanic*; the pearlite can be resolved in this image. The dark-gray areas are ferrite grains, and the very dark elliptical structure is an MnS particle identified by energy-dispersive X-ray (EDAX) analysis. It is elongated in the direction of the banding, suggesting that the banding was a result of hot-rolling the steel. Small nonmetallic inclusions can be seen, and some ferrite grain boundaries are also visible.

6.3.1   Weak Iron Rivets

Weak rivets [2, 4, 5] may have been the *Titanic*'s fatal flaw.

Metallurgists Tim Foecke and Jennifer Hooper McCarty found that the rivets holding together the *Titanic*'s steel plates toward the bow and the stern were made of low-grade iron. The hull plates and the rivets had different compositions. This resulted in galvanic corrosion, and the rivets corroded faster than the hull plates. The low-grade iron rivets would have ripped apart easily during the collision, causing the ship to sink faster than it would have had stronger rivets been used.

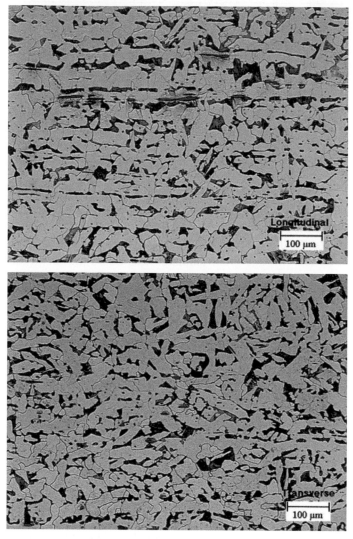

Figure 6.1 Optical micrographs of the *Titanic*'s hull steel in the (top) longitudinal and (bottom) transverse directions. The banding seen resulted in elongated pearlite colonies (which cannot be resolved in these micrographs) and MnS particles. Reprinted by permission from Springer Nature Customer Service Centre GmbH: Springer Nature, *JOM Journal of the Minerals, Metals and Materials Society*, Ref. [1], Copyright (1998).

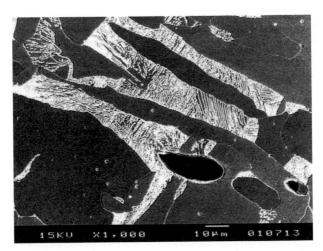

Figure 6.2 SEM micrograph of the polished and etched surface of steel from the *Titanic*. Reprinted by permission from Springer Nature Customer Service Centre GmbH: Springer Nature, *JOM Journal of the Minerals, Metals and Materials Society*, Ref. [1], Copyright (1998).

6.4 Tensile Testing

Each steel plate on *Titanic*'s hull was ~1.88 cm thick, while the bulkhead plate was 1.25 cm thick. Corrosion due to seawater had reduced the thickness of the hull plate, so it was impossible to take standard tensile specimens. A smaller tensile specimen with a reduced section of 0.625 cm diameter and 2.5 cm gauge length was used [6].

Table 6.3 Comparison of tensile testing of the *Titanic*'s steel and SAE 1020 steel [1]

	Titanic	**SAE 1020**
Yield strength	193.1 MPa	206.9 MPa
Tensile strength	417.1 MPa	379.2 MPa
Elongation	29%	26%
Reduction in area	57.1%	50%

Source: Reprinted by permission from Springer Nature Customer Service Centre GmbH: Springer Nature, *JOM Journal of the Minerals, Metals and Materials Society*, Ref. [1], Copyright (1998).

The tensile test results are given in Table 6.3 [1]. The data is compared with tensile test data for SAE 1020 steel, which is similar in composition. The table shows that the *Titanic's* steel had lower yield strength, possibly due to a larger grain size. The elongation increased as well, again due to the large grain size.

6.5   Charpy Impact Tests

Three series of standard Charpy specimens were used: the specimen axis parallel to the longitudinal direction in the *Titanic's* hull plate, the specimen axis parallel to the transverse direction, and a modern ASTM A36 steel series [6]. Charpy impact tests were performed on the specimens at a temperature ranging from –55°C to 179°C. The cleavage planes in ferrite were quite apparent, and cleavage plane surfaces at different levels were defined by straight lines. In addition, there were curved slip lines on the cleavage planes (Fig. 6.3). MnS particles were identified by EDAX analysis, some of which existed as protrusions from the surface. These protrusions were pulled out of the complementary fracture surface. In addition, intrusions remained after the MnS particles were pulled out of the fracture surface.

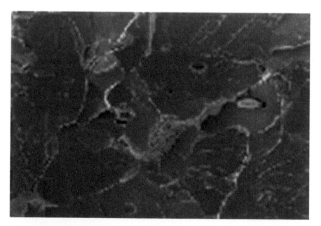

Figure 6.3 SEM micrograph of a Charpy impact fracture newly created at 0°C, showing cleavage planes containing ledges and protruding MnS particles. Reprinted by permission from Springer Nature Customer Service Centre GmbH: Springer Nature, *JOM Journal of the Minerals, Metals and Materials Society*, Ref. [1], Copyright (1998).

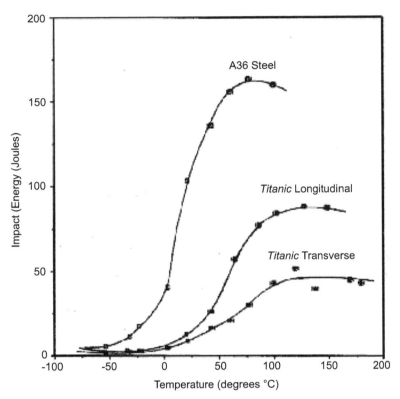

Figure 6.4 Charpy impact energy versus temperature for longitudinal and transverse specimens of the *Titanic's* hull plate and ASTM A36 steel. Reprinted by permission from Springer Nature Customer Service Centre GmbH: Springer Nature, *JOM Journal of the Minerals, Metals and Materials Society*, Ref. [1], Copyright (1998).

Figure 6.4 shows impact energy versus temperature curves for the three specimen series. At high temperature, the longitudinal hull plate specimens showed better impact properties than transverse specimens. At low temperature, the impact energy required to fracture the longitudinal and transverse specimens was similar. Severe banding was the reason for the differences between the impact energies required to cause fractures at high temperatures. ASTM A36 steel specimens had the best impact properties. The ductile–brittle transition temperature determined at an impact energy of 20 joules was –27°C for ASTM A36, 32°C for longitudinal hull plate specimens, and 56°C for transverse hull plate specimens

(Fig. 6.4). It is clear that the steel used for the hull was unsuitable for service at low temperatures.

6.6 Shear Fracture Percentage

At low temperatures, the impact energy that causes a fracture is low, a brittle fracture is observed. At high temperatures, where the impact energy that causes a fracture is high, a ductile fracture with a shear structure can be observed. Figure 6.5 shows the shear fracture percentage versus temperature. Using 50% of the shear fracture

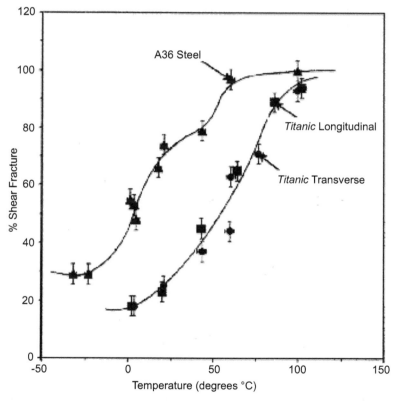

Figure 6.5 Shear fracture percentage from Charpy impact tests versus temperature for longitudinal and transverse specimens of the *Titanic*'s hull plate and ASTM A36 steel. Reprinted by permission from Springer Nature Customer Service Centre GmbH: Springer Nature, *JOM Journal of the Minerals, Metals and Materials Society*, Ref. [1], Copyright (1998).

area as a reference point, shear fracture would occur in ASTM A36 at –3°C and in the *Titanic* steel at 49°C in the longitudinal direction and at 59°C in the transverse direction. At high temperatures, the impact energies for the *Titanic*'s longitudinal steel specimens are substantially higher compared to the transverse steel specimens. The difference between the longitudinal and transverse shear fracture percentage is low.

6.7 Conclusions

The Titanic's hull was triple-riveted in the central three-fifths length with mild steel rivets and double-riveted in the bow and stern with wrought iron in order to ensure strength in the center where the maximum wave flex stresses were assumed to be located. Analysis of the steel rivets showed good strength, but the wrought iron rivets contained, on average, three times more slag in large pieces than optimal levels. These results indicated that the fabrication was done by inexperienced tradesmen (at the time the *Titanic* was constructed, wrought iron was made by hand).

References

1. Felkins, K., Leighly, Jr., H. P., and Jankovic, A. (1998). The Royal Mail Ship *Titanic*: did a metallurgical failure cause a night to remember? *JOM*, **50**(1), pp. 12–18.

2. Hackett, C., and Bedford, J. G. (1996). *The Sinking of the Titanic: Investigated by Modern Techniques* (The Northern Ireland Branch of the Institute of Marine Engineers and the Royal Institution of Naval Architects).

3. Ballard, R. B., and Archbold, R. (1987). *The Discovery of the Titanic* (New York: Warner Books).

4. Davies, R. (1995). *Historical Metallurgy*, **29**, p. 34.

5. Jankovic, A. (1991). *Did Metallurgy Sink the Titanic* (Senior Project Report, Department of Metallurgical Engineering, University of Washington, Seattle).

6. ASTM. (1995). *Standard Test Methods and Definitions for Mechanical Testing of Steel Products* (Philadelphia, PA: ASTM A370-95a), pp. 2, 7.

Chapter 7

The *Titanic* is Resting and Rusting

7.1 Where is the *Titanic* Now?

The Royal Mail Ship (RMS) *Titanic* broke apart into two pieces as she sank to the bottom of the Atlantic Ocean. Now, she is resting, and reacting with marine bacteria and being slowly converted into "rusticles," "rust flows," and "rust flakes." The destruction is being facilitated by iron-eating bacteria, such as *Halomonas titanicae* [1].

7.2 What is Happening to the *Titanic* Now?

Every day, the *Titanic* is undergoing corrosion due to its environment. Interesting new corrosion products are being produced in the deep sea [1].

7.2.1 Corrosion Products on the *Titanic*: Rusticles

Rusticles are believed to be the major corrosion product on the *Titanic*. They are considered similar to stalactites and can be tens of centimeters long. The outer surface of rusticles is smooth and red made up of iron oxyhydroxide. The core of rusticles is bright orange. "Goethite" [FeO(OH)] is the name given to the needle-like crystals constituting the core.

Titanic Corrosion
Susai Rajendran and Gurmeet Singh
Copyright © 2020 Jenny Stanford Publishing Pte. Ltd.
ISBN 978-981-4800-95-2 (Paperback), 978-1-003-00449-3 (eBook)
www.jennystanford.com

Figure 7.1 Rusticles covering the *Titanic*'s bow. Courtesy of NOAA/Institute for Exploration/University of Rhode Island (NOAA/IFE/URI) [2].

Figure 7.2 The partly collapsed bathroom of Captain Edward Smith, with the bathtub now filled with rusticles (http://oceanexplorer.noaa.gov/explorations/03titanic/media/titanic_bathtub.html) [2].

Figures 7.1 and 7.2 show the rusticles covering the *Titanic*'s bow and Captain Edward Smith's bathtub. Biological activity tests have confirmed that the rusticles grew in the presence of a sulfate-reducing species of bacteria that grow quickly under anaerobic (no oxygen) conditions [2].

7.2.2 Rust Flows and Rust Flakes

Other forms of corrosion products have also been found on the *Titanic*'s wreck. These include (i) rust flows and (ii) rust flakes.

Rust flows are present majorly on the ship's deck. Results of analyses have shown that rust flows are spreading at a rate of 10 cm per year.

Rust flakes are similar to rusticles in composition. Rust flakes are a mixture of two minerals, goethite and lepidocrite, and are formed in the presence of iron and bacteria.

7.3 Rust-Eating Bacterial Species Eating the *Titanic*

Corrosion is a natural, spontaneous, and thermodynamically stable process. In due course, the *Titanic* will finally be destroyed completely by nature. No one can prevent that. The corrosion of the ship is being enhanced by the presence of iron-eating bacteria. DNA technology has identified the primary bacterium eating into the *Titanic* as *Halomonas titanicae*, a gram-negative, halophilic species of proteobacteria. This species has been found on the rusticles covering the *Titanic*'s iron hull (Fig. 7.3). The bacterium is fast consuming the rust and the iron, too. Ultimately, the wreck will be converted into a dust of rust on the bottom of the Atlantic Ocean. One of the researchers, Dr. Henrietta Mann, has pointed out that the action of microorganisms, such as *H. titanicae*, will totally deteriorate the *Titanic* probably by 2030 [3–5]. In a few years' time, nothing will be left of the *Titanic*, except and websites and memories.

The wreck's condition is too poor to consider raising it. The only option is to try and salvage items from the wreck, which is what the company RMS Titanic, Inc., is doing for years to display them in a touring exhibit [6–8].

Figure 7.3 Rusticles hanging from the *Titanic*'s hull and containing *Halomonas titanicae* (http://oceanexplorer.noaa.gov/explorations/03titanic/rusticles/media/detached_rusticles.html) [2].

7.4 The Irish *Titanic* Historical Society

Ed Coghlan, chairman of the Irish *Titanic* Historical Society, said, "This research backs up what divers who have been down to the wreck have seen; that the ship is falling apart.

"Fortunately it has been photographed extensively and there are amazing videos to show us what it looked like underwater.

"In the future, people might think it is a shame we didn't do more to preserve it, but the truth is that to preserve it would cost an absolute fortune and is probably almost impossible.

"It may be that as the structure of the wreck disappears, more of the interior becomes accessible.

"We may be able to learn even more about *Titanic* once things like the mailbags, for example, become visible.

"*Titanic* is a very human story and it will be fascinating to see what happens to the wreck in the coming years."

7.5   MIC of the Gulf-of-Mexico Mooring Chain

A postinstallation inspection of a mooring system comprising polyester-and-chain at a depth of ~6000 feet provided proof of microbiologically influenced corrosion (MIC) in the form of rust tubercles, or rusticles. Rust tubercles are porous and are usually formed on steel wrecks underwater. The presence of rust tubercles indicates that underwater corrosion occurs in a low-oxygen environment.

In marine environments, iron-reducing bacteria and sulfate-reducing bacteria (SRB) cause corrosion. Miller et al. (2012) reported their results of analysis of the ambient conditions required for MIC to occur and compared rusticles found during the mooring inspection with those found on underwater shipwrecks, such as the *Titanic* [10]. The authors investigated of the type of iron used in mooring chains and the rate of rusticle formation and provide possible remedies to prevent rusticle growth on mooring chains [10].

7.6   Investigations of MIC of Submerged Steel and Wooden Shipwrecks, and the Role of Bioconcretions

MIC is correlated with SRB activities. The major product of the reaction is hydrogen sulfide (H_2S), which, in turn, leads to corrosion when it comes into contact with the steel surface. Microbiological investigations of MIC have reported the following sequence of events [11]:

1. Biomass attaches to the steel surface.
2. Biomass growth occurs.
3. Biomass growth affects the supporting matrices below the attachment surfaces
4. Matrices begin to corrode, leading to structural collapse.

Studies on mild steels, concrete, and wood have reported a pattern of microbiological activity that commonly occurs in the case of significant corrosion risks. The physical and chemical nature of the attachment surface, the nature of the environment at the biomass–surface interface, the nature of biomass growth, biomass–matrix interaction, and final states of stabilization or destruction of the matrix and associated biomass, all affect corrosion. The form and function of the active biomass interacting with the surfaces of the matrix are critical to MIC. MIC has been observed during investigation of corrosion of the steel used in the *Titanic* [11].

7.7 A Mössbauer Spectral Study of the Hull Steel and Rusticles Recovered from the *Titanic*

A 295 K conversion electron Mössbauer spectral study was conducted of the *Titanic*'s hull plate steel, oriented with the gamma-ray direction perpendicular or parallel to the microstructural banding.

The two spectra showed almost identical average magnetization orientations close to the plane of each hull plate. The hyperfine parameters were almost identical compared to α-iron. In addition, a 4.2–295 K transmission Mössbauer spectral study of the rusticles revealed small geothite particles undergoing superparamagnetic relaxation with a blocking temperature of ~300 K.

Approximately 2% of the Mössbauer spectral absorption area corresponded to a quadrupole doublet with hyperfine parameters that are typically found in green rust. The iron-containing components identified in the rusticles agreed with the powder X-ray diffraction (XRD) results that showed the presence of small poorly crystallized goethite particles and quartz and green rust traces. The results showed a normal size of 20 ± 5 nm for the goethite particles from both the average hyperfine field and broadening of the XRD peaks. The authors deduced the magnetic anisotropy constant of the goethite particles from the hyperfine field and particle size as 8 × 10^3 J/m^3 [12].

7.8 *Titanic*, the Anatomy of a Disaster: A Report from the Marine Forensic Panel (SD-7)

The tragedy of the *Titanic* is surrounded by many mysteries. To solve some of these mysteries, in August 1996, George Tulloch of RMS Titanic, Inc. organized a scientific investigation with the Discovery Channel. They wanted to answer questions such as:

- What was the extent of the damage caused by the iceberg?
- Did the ship break up into two or three pieces?
- Was the quality of the steel a factor in the tragedy?

The investigation used a sonar scan, metallurgical testing, and finite element analysis. The results confirmed that the area of damage on the ship was <13 ft^2. Also, the forces created by the flooding and the quality of steel caused the hull to rupture because of which the ship broke into three large pieces while sinking to the bottom of the Atlantic Ocean. In addition, a microbiology study of the wreck demonstrated that corrosion and microorganisms (some new, previously unknown to biologists) have eroded ~20% of the wreck [13].

7.9 Microworld of the *Titanic*

Discovery of the wreck of the *Titanic* has rekindled historical interest in the ship. The wreck of the *Titanic* also has considerable scientific value. Microscopic or geochemical investigations have revealed in the *Titanic* an active, complex world of microorganisms. The microorganisms have the power to maintain strong chemical gradients, which is confirmed by the presence of minerals that are stable at very different pH conditions within a small volume of rust [14].

7.10 Technically Plausible Scenario for Salvaging the *Titanic*

The wreck of the *Titanic* was discovered in 1985, 73 years after she sank. In 2000, in an expedition by RMS Titanic, Inc., 28 dives carried out and >800 artifacts recovered from the wreck, including the ship's engine telegraphs, perfume vials, and watertight door gears (Fig. 7.4) [2].

Figure 7.4 *DSV Alvin*, used in 1986 to mount the first manned expedition to the wreck of the *Titanic* [2]. Source: National Oceanic and Atmospheric Administration.

In 2003 and 2004, two expeditions to the *Titanic* were carried out by the US National Oceanic and Atmospheric Administration.

The first (June 22–July 2, 2003) performed four dives in two days to assess the wreck site's current condition and carry out scientific observations to support ongoing research. The explorers specifically targeted the stern section, which had previously received little attention, in addition to the microorganisms colonizing the wreck [2].

In the second expedition (May 27–June 12, 2004), Robert Ballard returned to the *Titanic* nearly 20 years after he discovered it. The expedition spent 11 days on the wreck to carry out high-resolution mapping using video and stereoscopic still images [2].

RMS Titanic, Inc., mounted further expeditions in 2004 and 2010 after the first comprehensive map of the *Titanic*'s entire debris field was created. The debris field measured 3 miles × 5 miles (4.8 km × 8.0 km), and the expedition used two torpedo-shaped robots to take sonar scans and >130,000 high-resolution images. Using these scans and images, scientists created a detailed photomosaic of the debris field in order to obtain a clear view of the dynamics of the *Titanic*'s sinking.

In 2012, Ballard announced a plan to preserve the *Titanic*'s wreck using deep-sea robots to clean and paint it with antifouling paint to preserve it in its current state forever.

Under the assumption that the national importance of salvaging the *Titanic* is greater compared to economic considerations, Neil Monney also proposed a technically plausible scenario for the project. The hull plate steel of the *Titanic* underwent much corrosion due to seawater, but this might have been inhibited by copper-rich sediments from hydrothermal sources.

The salvage operation described by Monney is beyond state of the art and the required technical capability could be developed in the next decade. In addition, alternative approaches need to be considered in order to refine the calculations for an actual salvage operation. Monney's approach is to "partially fill the hull of the *Titanic* with syntactic foam, reaching a condition of near-neutral buoyancy, and then applying the net lifting force with hydrogen-filled inflatable salvage pontoons that could provide a maximum net lifting force of 6000 LT" [15].

References

1. http://www.corrosion-doctors.org/Landmarks/titan-corrosion.htm

2. https://en.wikipedia.org/wiki/Wreck_of_the_RMS_Titanic (https://en.wikipedia.org/wiki/Wikipedia:Text_of_Creative_Commons_Attribution-ShareAlike_3.0_Unported_License)

3. Sánchez-Porro, C., Kaur, B., Mann, H. and Ventosa, A. (2010). Halomonas titanicae sp. nov., a halophilic bacterium isolated from the RMS Titanic. *IJSEM*, **60**(12), pp. 2768–2774.

4. Daily Mail Reporter. (10 December 2010). New rust-eating bacteria "destroying wreck of the Titanic". http://www.dailymail.co.uk/sciencetech/article-1336542/Halomonas-titanicae-New-rust-eating-bacteria-destroying-wreck-Titanic.html

5. Mason, B. (24 May 2011). Top 10 new species discovered in 2010. *Wired*. https://www.wired.com/2011/05/2010-species-gallery/

6. http://www.dailymail.co.uk/sciencetech/article-1346446/Titanic-wreck-completely-destroyed-20-years-new-rust-eating-bacteria.html

7. http://www.dailymail.co.uk/sciencetech/article-1346446/Titanic-wreck-completely-destroyed-20-years-new-rust-eating-bacteria.html#ixzz4b8BGAtm7

8. http://www.dailymail.co.uk/sciencetech/article-1346446/Titanic-wreck-completely-destroyed-20-years-new-rust-eating-bacteria.html#ixzz4b8Bfacla

9. BBC News. (6 December 2010). New species of bacteria found in Titanic "rusticles". http://www.bbc.com/news/science-environment-11926932

10. Miller, J.D., Warren, B.J. and Chabot, L.G. (2012). Microbiologically influenced corrosion of gulf of Mexico mooring chain at 6,000 feet depths. *Proceedings of the International Conference on Offshore Mechanics and Arctic Engineering - OMAE*, **2**, pp. 649–658.

11. https://www.researchgate.net/publication/292837622_Investigations_of_microbiologically_influenced_corrosion_MIC_of_submerged_steel_and_wooden_ship_wrecks_and_the_role_that_bio-concretions_play_in_these_degenerative_processes

12. Long, G.J., Hautot, D., Grandjean, F., Vandormael, D. and Leighly Jr., H.P. (2004). A Mössbauer spectral study of the hull steel and rusticles recovered from the Titanic. *Hyperfine Interactions*, **155**(1–4), pp. 1–13.

13. Garzke Jr., W.H., Brown, D.K., Matthias, P.K., et al. (1998). Titanic, the anatomy of a disaster: a report from the marine forensic panel (SD-7).

Transactions - Society of Naval Architects and Marine Engineers, **105**, pp. 3–48.

14. Stoffyn-Egli, P. and Buckley, D.E. (1995). Micro-world of the Titanic. *Chemistry in Britain*, **31**(7), pp. 551–553.

15. Monney, N.T. (1986). Technically plausible scenario for the salvage of the Titanic. *Ocean Science and Engineering*, **11**(3–4), pp. 115–135.

Chapter 8

Corrosion Inhibitors for Metals in an Acidic Medium

Rust and oxide layers on corroded metal are removed by the pickling process. To prevent loss of material during the pickling process, corrosion inhibitors are added. Extracts of natural products, thiourea, triazole derivatives, and sodium potassium tartrate have used as corrosion inhibitors. For evaluation of corrosion inhibition by inhibitors, various methods, such as the weight loss method, polarization study, and AC impedance spectra, have been employed. Protective films have been analyzed by various surface analysis techniques such as FTIR, SEM, EDAX, and AFM.

8.1 Pickling with Acids

Metal surfaces have to be cleaned before they are used for painting or designing of various parts of different types of vessels. For this purpose, metals with rust and an oxide surface are treated with concentrated acids, such as hydrochloric acid or sulfuric acid. This process is called pickling. During this process, the base metal decays. To avoid this unavoidable corrosion process, inhibitors are added. In this chapter, inhibitors used in an acidic medium and the methods involved in corrosion inhibition studies are discussed.

Titanic Corrosion
Susai Rajendran and Gurmeet Singh
Copyright © 2020 Jenny Stanford Publishing Pte. Ltd.
ISBN 978-981-4800-95-2 (Paperback), 978-1-003-00449-3 (eBook)
www.jennystanford.com

8.1.1 Metals

Many inhibitors have been used to control corrosion of various metals, such as mild steel [1, 7, 17], aluminum and its alloys [5, 8], and stainless steel [2, 6], in an acidic medium.

8.1.2 Medium

For an acidic medium, hydrochloric acid [6, 7, 11] and sulfuric acid [5] are used. Mainly they are used for the pickling process.

8.1.3 Temperature

The experiments have been conducted at various temperatures [1, 12].

8.1.4 Inhibitors

To prevent corrosion of metals in an acidic medium, inhibitors like extracts of natural products [1, 5, 7, 14, 16, 18], quinolline [3], organic inhibitors [4], thiourea derivatives [6, 10, 15], citric acid [8], triazole derivatives [9], and succinic acid derivatives [12] have been used.

8.1.5 Methods

To evaluate the corrosion inhibition efficiency of inhibitors, several methods such as the weight loss method [1, 2, 5] and electrochemical studies [1, 5, 7] such as polarization and AC impedance spectra have been employed.

8.1.6 Surface Morphology of the Protective Film

The protective films formed on metal surfaces have been analyzed by various surface analysis techniques like Fourier transform infrared spectroscopy (FTIR) [5, 13], scanning electron microscopy (SEM) [1, 5, 8], energy-dispersive X-ray analysis (EDAX) [20], atomic force microscopy (AFM) [16], X-ray photoelectron spectroscopy (XPS) [8], and nuclear magnetic resonance (NMR) [13].

8.1.7 Adsorption Isotherm

The protective film is formed on the metal surface by adsorption of inhibitors on the metal surface. The adsorption isotherms are usually Temkin adsorption isotherm [6, 15] and Langmuir adsorption isotherm [13,14, 20].

8.1.8 Thermodynamic Parameters

From the adsorption isotherms, various thermodynamic parameters such as changes in free energy, entropy, enthalpy, and activation energy [1, 12] have been calculated.

Table 8.1 summarizes various inhibitors, methods used to evaluate corrosion inhibition efficiencies of inhibitors, and various surface analysis techniques.

8.2 Case Study: Corrosion Resistance of Mild Steel Immersed in 1N HCl in the Absence and Presence of Sodium Potassium Tartrate

Corrosion resistance of mild steel immersed in 1N HCl in the absence and presence of sodium potassium tartrate (SPT) has been evaluated by electrochemical studies such as polarization studies and AC impedance spectra. Electrochemical studies were carried out in a CHI 660 A workstation. A three-electrode cell assembly was used. Mild steel was the working electrode. Platinum foil was the counterelectrode. A saturated calomel electrode was the reference electrode.

8.2.1 Analysis of Results of Polarization Studies

Corrosion resistance of mild steel immersed in 1N HCl in the absence and presence of SPT has been evaluated by polarization studies. Polarization curves are shown in Fig. 8.1. The corrosion parameters such as corrosion potential, Tafel slopes (b_c = cathodic; b_a = anodic), linear polarization resistance (LPR), and corrosion current (I_{corr}) are given in Table 8.2. It is well known that when corrosion resistance increases, the LPR value increases and the corrosion current value decreases.

Table 8.1 Inhibitors for an acidic medium

S. no.	Metal/Medium	Inhibitor/ Additives	Methods	Findings	Refs.
1	Mild steel and copper metals in both 0.5 M HCl and 0.5 M NaOH media	Tender areca nut husk (green color) extract	Weight loss measurements, SEM Tafel plot and AC impedance studies Temperature studies	Corrosion protection efficiencies were enhanced with increment in the concentration of plant extracts.	1
2	Superheater tube, ferritic steel, and austenitic stainless steel such as 12Cr1MoV, T22, T91, TP347, and SUS304, acid	Inhibitor	Weight loss method to evaluate the corrosion rate	The traditional weight loss method to evaluate the corrosion rate of materials in the chemical-cleaning process is discussed to monitor the corrosion process.	2
3	Different metals, acidic media	Quinoline derivatives	Review paper	These inhibitors slow down the corrosion rate and thus prevent economic losses due to metallic corrosion on industrial vessels, equipment, or surfaces.	3

S. no.	Metal/Medium	Inhibitor/ Additives	Methods	Findings	Refs.
4	Ferrous and nonferrous metals in acidic solutions	Organic inhibitors	Review paper	Heteroatoms of these molecules exist in the form of polar functional groups, which can actively participate in adsorption processes, in addition to the π-electrons of multiple bonds.	4
5	Al–Si–Fe/SiC composite in sulfuric acid medium	*Acacia senegalensis* stem extract	Weight loss, gravimetric and potentiostatic polarization techniques, FTIR and GC–gas chromatography–mass spectrometry (GC-MS), SEM	High inhibition efficiency was obtained. The *A. senegalensis* stem extract functions as an effective inhibitor of the composite in 0.5 M sulfuric acid medium.	5
6	304 SS in HCl medium	*N*-napthyl-*N*l-phenyl thiourea	Tafel extrapolation and impedance spectroscopy methods	The compound follows Temkin's isotherm model, and the inhibition is governed by physical adsorption along with weak chemical bonding.	6
7	Mild steel corrosion in an aggressive HCl medium	*Sesbania grandiflora* leaf extract	Weight loss investigation, open-circuit voltage analysis, Tafel polarization, AC impedance analysis, infrared (IR) spectroscopy, and energy-dispersive X-ray spectroscopy analysis	*S. grandiflora* leaf extract can be used as economical and eco-friendly inhibitor to prevent corrosion of mild steel in HCl medium.	7

(Continued)

Table 8.1 (*Continued*)

S. no.	Metal/Medium	Inhibitor/Additives	Methods	Findings	Refs.
8	Aluminum alloys in alkaline media	Citric acid	Impedance measurements via dynamic electrochemical impedance spectroscopy (EIS) in galvanostatic mode (g-DEIS), electrochemical, XPS, and SEM studies	The conclusion was reached that regardless of corrosion resistance, the kinetics and mechanism of passive layer formation in Al alloys depend only slightly on alloying additives. The AA7020 alloy was an exception because of a specific interaction between zinc and citric acid.	8
9	Metals and alloys in acid/base medium	1,2,4-Triazole derivatives as corrosion inhibitors		1,2,4-Triazole derivatives can be used as corrosion inhibitors to prevent corrosion of metals and alloys in acid/base medium.	9
10	Mild steel in acidic medium	N-benzyl-N′-phenyl thiourea (BPTU) and N-cyclohexyl-N′-phenyl thiourea (CPTU)	Tafel extrapolation technique, Temkin's adsorption isotherm	Governed by a chemisorption mechanism, this shows that there was a good correlation between the Tafel extrapolation and linear polarization results.	10

S. no.	Metal/Medium	Inhibitor/Additives	Methods	Findings	Refs.
11	Mild steel in 4% hydrochloric acid	Liquorice (*Glycyrrhiza glabra*)	Weight loss and electrochemical experiment	The efficiency attained was more than 62.4% at 0.5% of inhibitor in the weight loss experiment.	11
12	Mild steel in 0.5M NaCl	2-[2-Methyl-4(or 5)-alkylisoxazolidin-5(or 4)-yl)methyl] succinic acids	Gravimetric and electrochemical methods, EIS	The adsorption of inhibitors obeyed Temkin adsorption isotherm model. The inhibitors were adsorbed on metal surface by chemisorption process. The XPS study confirmed the adsorption of the inhibitors on the metal surface.	12
13	Aluminum alloy composite in acidic media	Ionic liquid, 1.3-bis[2-(4-methoxyphenyl)-2-oxoethyl]-1H-benzimidazol-3-ium bromide (MOBB)	FTIR, NMR, liquid chromatography–mass spectrometry (LC-MS), and single-crystal XRD (SCXRD) techniques, EIS, and potentiodynamic polarization methods	Adsorption of inhibitors on the metal surface obeyed the Langmuir adsorption isotherm. The adsorption process was mainly through the chemisorption process.	13
14	Monel 400 alloy in acidic media	Green leaves of *Mespilus japonica*, *Ricinus communis L.*, and *Vitis vinifera*	EIS and potentiodynamic polarization	Adsorption of inhibitors on the metal surface obeyed Langmuir adsorption isotherm. These inhibitors are eco-friendly inhibitors.	14

(*Continued*)

Table 8.1 (*Continued*)

S. no.	Metal/Medium	Inhibitor/ Additives	Methods	Findings	Refs.
15	304 SS in 1M HCl solution	*N*-furfuryl-*N*l-phenyl thiourea (FPTU)	Tafel extrapolation and EIS techniques	Adsorption of inhibitors on the metal surface obeyed Temkin's adsorption isotherm. The inhibitor controlled corrosion of 304 SS in 1 M HCl solution.	15
16	Mild steel in chloride media	Ethanol extract of propolis	Classical electrochemical and electrochemical probe beam deflection techniques, FTIR, high-performance liquid chromatography, SEM, AFM	This work is pioneering in its evaluation of the inhibitory capacity of a natural product whose main constituent is 3,5-diprenyl-4-hydroxycinnamic acid (DHCA).	16
17	D16T aluminum alloy	Rhamnolipid biocomplex	Adsorption process	The biosurfactants added to the corrosive medium increase the rate of recovery of protective films on the aluminum alloy in the stage of repassivation by a factor of 2–4, as compared with the uninhibited medium.	17

S. no.	Metal/Medium	Inhibitor/ Additives	Methods	Findings	Refs.
18	Metals and alloys in acid pickling processes	Green corrosion inhibitors,		Natural products have been used as eco-friendly corrosion inhibitors for various metals and alloys in different acid media. They offer good corrosion inhibition efficiencies. A mechanism of corrosion inhibition has been proposed.	18
19	Carbon steels and high alloys in 4M hydrochloric acid	Polymerizable organic corrosion inhibitors	Weight loss technique	The influence of metal composition on the efficiency of film-forming corrosion inhibitors and the synergetic behavior with quaternary ammonium cations are discussed.	19
20	6061 Aluminum alloy in 0.25M hydrochloric acid solution	Biopolymer starch as an ecofriendly green inhibitor	Potentiodynamic polarization and EIS techniques, SEM-EDX analysis	The inhibitor molecules are chemisorbed on the metal surface. It follows the Langmuir adsorption isotherm.	20

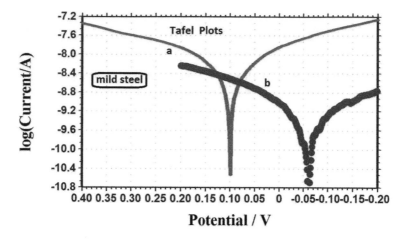

Figure 8.1 Polarization curves of mild steel in various test solutions: (a) 1N HCl and (b) 1N HCl + SPT (500 ppm).

It is observed from Table 8.2 that in the presence of an inhibitor, the LPR value increases (Fig. 8.2) and the corrosion current value decreases. This indicates that SPT is a very good corrosion inhibitor for mild steel in 1N HCl solution.

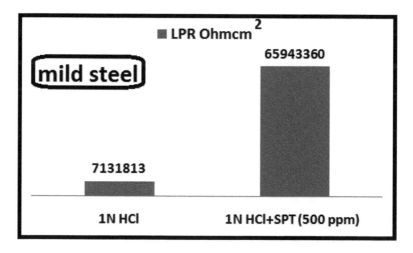

Figure 8.2 Comparison of LPR values.

Table 8.2 Corrosion parameters of mild steel immersed in 1N HCl in the absence and presence of SPT (500 ppm) obtained by polarization study

System	E_{corr} (mV$_{SCE}$)	b_c (mV/ decade)	b_a (mV/ decade)	LPR (Ohm cm^2)	I_{corr} (A/cm^2)
1N HCl	−99	1/4.85	1/4.761	7131813	6.343×10^{-9}
1N HCl+SPT (500 ppm)	−61	1/4.43	1/5.784	65943360	6.455×10^{-10}

8.2.2 Analysis of Results of AC Impedance Spectra

Corrosion resistance of mild steel immersed in 1N HCl in the absence and presence of SPT has also been evaluated by AC impedance spectra (Figs. 8.3–8.12). The corrosion parameters such as charge transfer resistance (R_t), impedance value (log[z/ohm]), double-layer capacitance (C_{dl}), and phase angle values are given in Table 8.3.

It is well known that when corrosion resistance increases, the charge transfer resistance value increases and the double-layer capacitance value decreases.

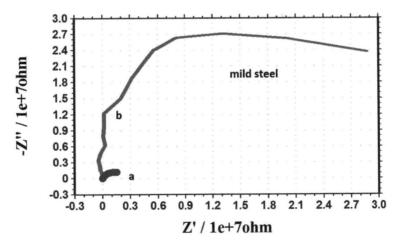

Figure 8.3 Nyquist plots of mild steel in various test solutions: (a) 1N HCl and (b) 1N HCl + SPT (500 ppm).

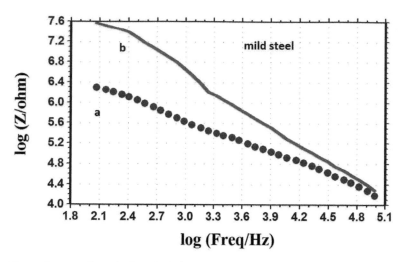

Figure 8.4   Log(Freq/Hz) vs. log(Z/ohm) plot.

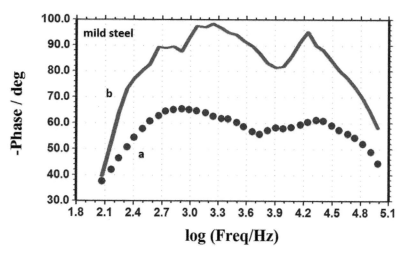

Figure 8.5   Log(Freq/Hz) vsm –phase/deg plot.

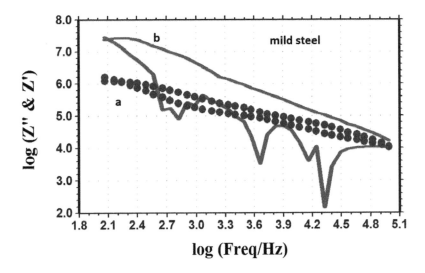

Figure 8.6   Log(Freq/Hz) vs. log(Z'' & Z')plot.

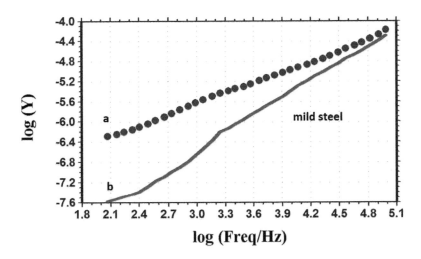

Figure 8.7   Log(Freq/Hz) vs. log(Y) plot.

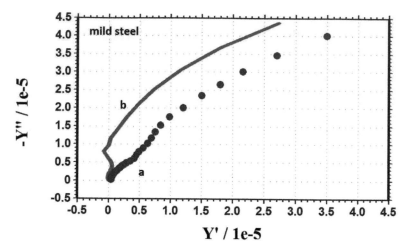

Figure 8.8 Y' vs. $-Y''$ plot.

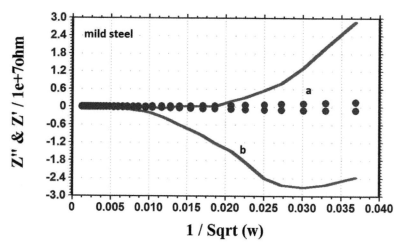

Figure 8.9   1/Sqrt(w) vs. Z'' & Z' plot.

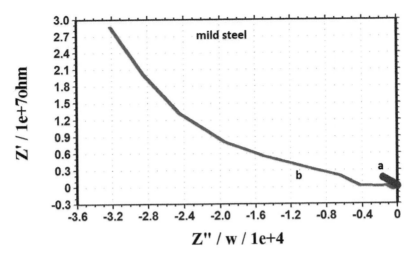

Figure 8.10 Z''/w vs. Z' plot.

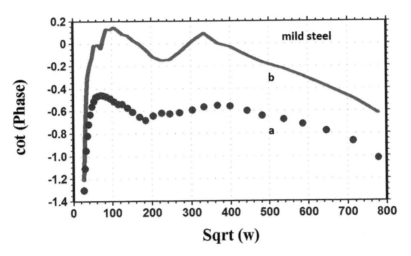

Figure 8.11 Sqrt(w) vs. cot(phase) plot.

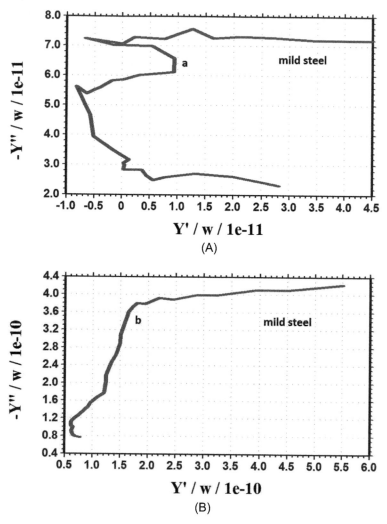

Figure 8.12   (A) *Y′/w* vs. −*Y″/w* plot. (a) 1N HCl. (B) *Y′/w* vs. −*Y″/w* plot. (b) 1N HCl + SPT (500 ppm).

It is observed from Table 8.3 that in the presence of an inhibitor, the charge transfer resistance value increases (Fig. 8.13) and the double-layer capacitance value decreases. This indicates that SPT is a very good corrosion inhibitor for mild steel in 1N HCl solution.

Table 8.3 Corrosion parameters of mild steel immersed in 1N HCl in the absence and presence of SPT (500 ppm) obtained by AC impedance spectra

System	R_t (Ohm cm^2)	C_{dl} (F/cm^2)	Impedance Log(z/ohm)	Phase angle (°)
1N HCl	1537840	3.316×10^{-8}	6.3	65.34
1N HCl+SPT (500 ppm)	28680000	1.778×10^{-9}	7.559	98.62

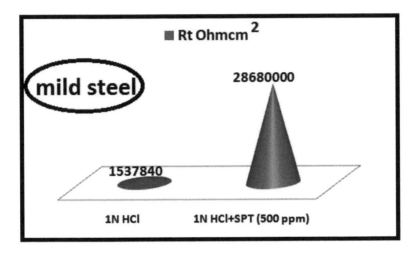

Figure 8.13 Comparison of R_t values.

References

1. Raghavendra, N., and Ishwara Bhat, J. (2019). Application of green products for industrially important materials protection: an amusing anticorrosive behavior of tender arecanut husk (green color) extract at metal-test solution interface, *Measurement*, **135**, pp. 625–639.

2. Li, H., Xue, Q., Nie, X., and Xu, Y. (2018). Investigation of chemical cleaning of supercritical superheater oxide scale, *MATEC Web Conf.*, **238**, p. 02011.

3. Lavanya, K., Saranya, J., and Chitra, S. (2018). Recent reviews on quinoline derivatives as corrosion inhibitors, *Corros. Rev.*, **36**(4), pp. 365–371.

4. Goyal, M., Kumar, S., Bahadur, I., Verma, C., and Ebenso, E. E. (2018). Organic corrosion inhibitors for industrial cleaning of ferrous and non-ferrous metals in acidic solutions: a review, *J. Mol. Liq.*, **256**, pp. 565–573.

5. Suleiman, I. Y., Abdulwahab, M., and Sirajo, M. Z. (2018). Anti-corrosion properties of ethanol extract of acacia senegalensis stem on Al–Si–Fe/ SiC composite in sulfuric acid medium, *J. Fail. Anal. Prev.*, **18**(1), pp. 212–220.

6. Shetty, N., and Herle, R. (2018). Inhibiting action of N-Napthyl-Nl-Phenyl thiourea on the corrosion of 304 Ss In HCL medium, *Int. J. Civ. Eng. Technol.*, **9**(2), pp. 507–515.

7. Krishnan, A., and Shibli, S. M. A. (2018). Optimization of an efficient, economic and eco-friendly inhibitor based on Sesbania grandiflora leaf extract for the mild steel corrosion in aggressive HCl environment, *Anti-Corros. Methods Mater.*, **65**(2), pp. 210–216.

8. Wysocka, J., Krakowiak, S., and Ryl, J. (2017). Evaluation of citric acid corrosion inhibition efficiency and passivation kinetics for aluminium alloys in alkaline media by means of dynamic impedance monitoring, *Electrochim. Acta*, **258**, pp. 1463–1475.

9. Phadke Swathi, N., Alva, V. D. P., and Samshuddin, S. (2017). A review on 1,2,4-triazole derivatives as corrosion inhibitors, *J. Bio- Tribo-Corros.*, **3**(4), p. 42.

10. Shetty, S. D., and Shetty, N. (2017). Inhibition of mild steel corrosion in acid medium, *Int. J. Technol.*, **8**(5), pp. 909–919.

11. Koolivand Salooki, M., Esfandyari, M., and Koulivand, M. (2017). Inhibition corrosion by liquorices, *J. Corros. Sci. Eng.*, **20**.

12. Ali, S. A., Mazumder, M. A. J., Nazal, M. K., and Al-Muallem, H. A. (2017). Assembly of succinic acid and isoxazolidine motifs in a single entity to mitigate CO_2 corrosion of mild steel in saline media, *Arabian J. Chem.*, In Press.

13. Kshama Shetty, S., and Nityananda Shetty, A. (2017). Eco-friendly benzimidazolium based ionic liquid as a corrosion inhibitor for aluminum alloy composite in acidic media, *J. Mol. Liq.*, **225**, pp. 426–438.

14. Kherraf, S., Zouaoui, E., and Medjram, M. S. (2017). Corrosion inhibition of Monel 400 in hydrochloric solution by some green leaves, *Anti-Corros. Methods Mater.*, **64**(3), pp. 347–354.

15. Divakara Shetty, S., Herle, R., and Shetty, N. (2017). Experimental study on corrosion inhibition of 304 SS in acid solution, *Int. J. Appl. Eng. Res.*, **12**(22), pp. 12892–12898.

16. Dolabella, L. M. P., Oliveira, J. G., Lins, V., Matencio, T., and Vasconcelos, W. L. (2016). Ethanol extract of propolis as a protective coating for mild steel in chloride media, *J. Coat. Technol. Res.*, **13**(3), pp. 543–555.

17. Zin, I. M., Karpenko, O. V., Kornii, S. A., et al. (2016). Influence of a rhamnolipid biocomplex on the corrosion of duralumin in the case of mechanical activation of its surface, *Mater. Sci.*, **51**(5), pp. 618–626.

18. Khan, G., Newaz, K. M. S., Basirun, W. J., et al. (2015). Application of natural product extracts as green corrosion inhibitors for metals and alloys in acid pickling processes: a review, *Int. J. Electrochem. Sci.*, **10**(8), pp. 6120–6134.

19. Barmatov, E., Hughes, T., and Nagl, M. (2015). Performance of organic corrosion inhibitors on carbon steels and high alloys in 4M hydrochloric acid, *NACE - International Corrosion Conference Series*, NACE-2015-5893.

20. Charitha, B. P., and Rao, P. (2015). Ecofriendly biopolymer as green inhibitor for corrosion control of 6061-aluminium alloy in hydrochloric acid medium, *Int. J. ChemTech Res.*, **8**(11), pp. 330–342.

Chapter 9

Inhibitors for Rebars in Simulated Concrete Pore Solution: An Overview

Reinforced concrete plays a major role in the modern construction field, and the degradation of concrete structures due to corrosion in reinforcing steel results in damage to people and property. The reinforcing steel in concrete can be protected from corrosion by the formation of a passive film on its surface in a concrete pore solution (pH 12.5–13.5), and this passive film can be damaged due to pH or an increase in chloride concentration at the steel–concrete interface. The corrosion behavior of steel in a simulated concrete pore solution (SCPS) is analyzed in the absence and presence of organic, inorganic, and natural extracts. The corrosion resistance property of inhibitors is analyzed by the weight loss method and various electrochemical methods. The surface morphology has been analyzed by UV, NMR, FTIR, XPS, SEM, EDAX, and AFM.

9.1 Introduction

Reinforced concrete (RC) is a complex substance—concrete in which steel is implanted in such a manner as to absorb the tensile, shear, and compressive stresses in the concrete structure. For strong, durable concrete structures, the reinforcement needs to have high toleration to tensile strain, elevated relative strength, and a good bond with the concrete irrespective of moisture, changing

Titanic Corrosion
Susai Rajendran and Gurmeet Singh
Copyright © 2020 Jenny Stanford Publishing Pte. Ltd.
ISBN 978-981-4800-95-2 (Paperback), 978-1-003-00449-3 (eBook)
www.jennystanford.com

temperature, and pH. Reinforcing steel in concrete can be protected from corrosion by forming a compressed passive film on its surface with a concrete pore solution with high alkalinity (pH 12.5–13.5). The corrosion behavior of reinforcing steel in a simulated concrete pore solution (SCPS) is analyzed in the absence and presence of various types of inhibitors. SCPS mainly consists of saturated calcium hydroxide $Ca(OH)_2$, sodium hydroxide (NaOH), and potassium hydroxide (KOH) with the pH 13.5 [1, 2].

Various inhibitors are used to prevent corrosion in steel immersed in SCPS in various mediums and at different temperatures and pH values. The corrosion resistance property of inhibitors is analyzed by the weight loss method and electrochemical studies such as polarization study, AC impedance spectra, surface analysis by UV absorption spectra, fluorescence spectra, Fourier transform infrared spectroscopy (FTIR), scanning electron microscopy (SEM), and atomic force microscopy (AFM) [3–10].

9.2 Metals

Corrosion resistance of various metals in concrete solution, such as mild steel [13–15], galvanized steel [17], prestressed steel [52, 74], polypyrolecoated steel [25], carbon steel, steel [34, 36, 37, 43, 44, 46–48, 50, 51, 58–63, 66–68, 71–73, 75], has been investigated.

9.3 Inhibitors

Among various inhibitors that have been used along with concrete admixtures, mention may be made of D-sodium gluconate [29, 41, 50, 70, 72,], sodium nitrate [14, 29, 62, 63, 67], natural extracts [40, 64], and phosphate [34, 36, 57, 62, 75], which have been used to control the corrosion of various metals and their alloys in various mediums.

9.4 Methods

Various methods have been used to evaluate the inhibition efficiency (IE) of corrosion inhibitors in SCPS. Usually the weight loss method [14, 18, 31, 32, 42, 48, 50, 61, 63, 67, 68], cyclic voltammetry (CV) [35, 49], polarization study [11, 12, 18–21, 24, 25, 27–30, 32–35, 37,

38, 41, 45, 46, 49–51, 53, 55–57, 59, 61, 63–65, 67, 70–73, 75–77], and electrochemical impedance spectroscopy (EIS) [11, 12, 16, 17, 19, 20, 24, 28, 29, 33–35, 41, 49, 51, 53, 55, 62, 72, 75, 77] have been used.

9.5 Surface Analysis

The protective film formed on the metal surface during the process of corrosion protection of various metals by inhibitors has been analyzed by various surface analysis techniques such as UV absorption spectra [21], microscopic studies [14, 15, 22, 30, 31], FTIR [11, 12, 14, 31, 40, 42, 45, 46, 75], nuclear magnetic resonance (NMR) [40], SEM [11–13, 17, 19, 22, 27, 32, 34, 37, 38, 40, 42, 49, 70, 75], X-ray diffraction (XRD) [11, 12, 22, 26, 32, 40, 45, 46, 49, 54, 57, 58, 65, 75, 78], energy-dispersive X-ray spectroscopy (EDX) [15, 37, 42], and AFM [40, 45]. In general it has been observed that the protective film [13, 26, 41, 42, 56–58, 71–73] consists of a metal–inhibitor complex along with $CaCO_3$ and $Ca(OH)_2$ and the absorption of inhibitor on the metal surface is explained by adsorption isotherms [31, 33, 55, 64, 66].

9.6 Conclusion

Corrosion behavior of various steel alloys in SCPS has been investigated in the presence and absence of inorganic and organic inhibitors. Usually mild steel, prestressed steel, galvanized steel, polycoated steel, and steel rebar have been used. Organic inhibitors, inorganic inhibitors, and natural products are used as inhibitors. Corrosion resistance of metals has been evaluated by the weight loss method, electrochemical studies such as polarization study, and AC impedance spectra, and the protective film formed on the metal surface has been analyzed by SEM, FTIR, X-ray photoelectron spectroscopy (XPS), AFM, and EDX. The protective film consists of a metal–inhibitor complex, calcium carbonate, and calcium hydroxide. Experiments can be carried out at room temperature and at various pH levels. Corrosion inhibitors form a protective layer on the steel surface and prevent corrosion. The metals used in SCPS, various methods employed, inhibitors used, and important findings by various researchers are summarized in Table 9.1.

Table 9.1 Inhibitors used in SCPS for corrosion resistance

S. no.	Metals used/ Medium	Inhibitor/Additive	Method	Findings	Ref.
1	Steel/SCPS	Nitrate, nitrite-intercalated Mg–Al layered double hydroxides	EIS in the solution, SEM, XRD, and infrared spectroscopy	Corrosion of steel for LDH-NO_2 has a higher chloride threshold level. The dual effects, the uptake of chloride ions as an absorbent, and the release of NO_2- ions as an inhibitor contribute to better inhibition effect for LDH-NO_2.	11
2	Steel/SCPS	Polyvinyl Pyrrolidone (PVP)	EIS and polarization curve measurements, SEM, XPS, and Raman spectroscopy	PVP with the optimum concentration could effectively inhibit corrosion of the reinforcing steel and protect the steel from corrosion.	12
3	carbonsteel/ SCPS	Diamino((2-((2-aminoethyl)amino)ethyl) amino)methanethiol	Potentiodynamic polarization curves, EIS, and SEM	This reduced both anodic and cathodic current densities by forming a protective film on its surface.	13
4	Mild steel/ SCPS	Sodium nitrite and sodium citrate/zinc acetate	Weight loss, electrochemical studies, FTIR, XRD, microscopic studies	In the presence of zinc acetate, corrosion protection efficiency of trisosium citrate (TSC) increased appreciably.	14

S. no.	Metals used/ Medium	Inhibitor/Additive	Method	Findings	Ref.
5	Mild steel/ SCPS	$Na_3PO_4 \cdot 12H_2O$	Acoustic emission (AE), optical microscope (MO) and EDX	Phosphate ions help to delay prestressed steel corrosion initiation.	15
6	Mild steel/ SCPS	Chamaerops humilis	Open-circuit potential (OCP) and EIS	The extract plays an important role in the corrosion resistance potential.	16
7	Galvanized steel/SCPS	Sodium molybdate	OCP and EIS, SEM, and XPS	This acted as a barrier effect to suppress rapid dissolution.	17
8	Carbon steel/ SCPS	*N,N'*-dimethylaminoethanol	Weight loss and electrochemical techniques	Formation of a passive layer on the metal surface separates the metal from the corrosive medium.	18
9	Mild steel/ SCPS	Dioctyl sebacate (DS)	Potentiodynamic polarization, EIS diagrams, Mott–Schottky curves, and SEM	Corrosion IE of DS is superior.	19
10	Mild steel/ SCPS	DNA	Linear polarization resistance, EIS, and potentiodynamic polarization	DNA formed a dense film on the steel surface, which could effectively restrict the corrosion attack.	20
11	Mild steel/ SCPS	Poly(ethylene oxide)-b-polystyrene	UV spectrophotometry	–	21
12	Mild steel/ SCPS	Calcium lignosulfonate and sodium oleate	Microscopic infrared imaging	–	22

(Continued)

Table 9.1 (*Continued*)

S. no.	Metals used/ Medium	Inhibitor/Additive	Method	Findings	Ref.
13	Carbon steel/ SCPS	Sodium oleate	Potentiodynamic polarization, EIS, SEM, and XPS	Sodium oleate offers good corrosion protection to carbon steel immersed is simulated concrete pore solution. This may find application in concrete technology.	23
14	Carbon steel/ SCPS	Carboxylate of benzoic acid and dimethylethanolamine	Linear polarization resistance and EIS	Carboxylate corrosion inhibitors improve the chloride threshold level for carbon steel and decrease the corrosion rate (CR).	24
15	Polypyrrole coated steel / SCPS	Polypyrrole	Electrochemical polarization	It is possible to use a conducting polypyrrole-based paint-coating system for the protection of rebar in concrete.	25
16	Steel/SCPS	Cetyltrimethylammonium bromide (CTMAB)	Electrochemical method and XPS	CTMAB acted as an adsorption-type inhibitor and improved the corrosion resistance of the steel by forming an adsorption film on the steel surface.	26

S. no.	Metals used/ Medium	Inhibitor/Additive	Method	Findings	Ref.
17	Steel/SCPS	$NaNO_2$ and Na_2SiO_3	Scanning microelectrode technique assisted by polarization curve measurements and surface analyses	$NaNO_2$ with Na_2SiO_3 could effectively protect the reinforcing steel from corrosion.	27
18	Steel/SCPS	Molybdate ions	Potentiodynamic polarization, and EIS	The sodium molybdate inhibitor promotes the formation of a stable passive layer.	28
19	Carbon steel (UNS G10200)/ SCPS	Sodium D-gluconate, benzotriazole, sodium nitrite	Potential polarization and EIS	The corrosion process decreases.	29
20	Steel/SCPS	Na_2MoO_4	Potentiodynamic polarization measurements, micrographs, and XPS	These prevent corrosion of reinforcing steel.	30
21	Carbon steel/ SCPS	Calcium lignosulfonate (CLS)	Weight loss, polarization, fluorescence microscopy (FM), SEM microscopic infrared spectral imaging (M-IR), and XPS	The adsorption of CLS on the carbon steel surface occurred probably by both physisorption and chemisorption.	31

(Continued)

Table 9.1 (*Continued*)

S. no.	Metals used/ Medium	Inhibitor/Additive	Method	Findings	Ref.
22	Mild steel/ SCPS	Biomolecules	Weight loss, polarization, FM, SEM M-IR, and XPS	Biomolecules seem to be a film-forming mixed-type inhibitor.	32
23	Carbon steel/ SCPS	Aryl aminoalcohols:*N,N*-dimethylaminoethanol (DMEA)	Electrochemical polarization and EIS	Mixed inhibitor, Langmuir's adsorption isotherm.	33
24	Steel/SCPS	Trisodium phosphate	Potentiodynamic polarization, EIS, and SEM	Formation of a passive layer rich in phosphate on the steel surface.	34
25	Carbon steel/ SCPS	Polyhydroxy C-glycosidic ketone	Potentiodynamic polarization, EIS, and CV	The inhibitor exhibited excellent chemical stability and IE in a strong alkaline solution.	35
26	Steel/SCPS	Phosphate	The inhibitor exhibited excellent chemical stability and IE in strong alkaline solution.	Phosphate ions behaved as mixed-type corrosion inhibitors.	36
27	Steel/SCPS	Aspartic and lactic acid	Linear potentiodynamic polarization and gravimetric method, SEM/EDX	These promote the formation of a massive scale, reducing corrosion propagation.	37

S. no.	Metals used/ Medium	Inhibitor/Additive	Method	Findings	Ref.
28	Mild steel/ SCPS	–	Potentiodynamic polarization and SEM	Increase in the sulfate concentration from 0 to 2000 ppm increased the CR.	38
29	Mild steel/ SCPS	Mezlocillin (MZN) extracts	Reflectance spectra	Formation of an adsorbed film of yjr inhibitor on reinforcing steel in SCPS.	39
30	Mild steel/ SCPS	Extracts of natural products	NMR, FTIR spectra, SEM, AFM, and XRD	–	40
31	Mild steel/ SCPS	D-sodium gluconate	Electrochemical techniques, corrosion potential, potentiodynamic polarization, and EIS	Formation of a protective film on metallic surfaces.	41
32	20SiMn steel/ SCPS	Phytic acid	Weight loss and electrochemical methods. FTIR, reverse-phase high-performance liquid chromatography (RP-HPLC), SEM, and EDS	A film forms on the surface of the specimen.	42
33	Steel/SCPS	Amino alcohol and lithium nitrite	Autoclave accelerated corrosion test	The amino alcohol inhibitor also displayed anticorrosion properties.	43

(Continued)

Table 9.1 (*Continued*)

S. no.	Metals used/ Medium	Inhibitor/Additive	Method	Findings	Ref.
34	Q235 steel/ SCPS	*N,N*-dimethyl ethanolamine, diethanolamine, propylamine, ethanolamine	Quantum chemistry calculation, molecular dynamics simulation, and electrochemical test	Protective films were formed by the inhibitors.	44
35	C-steel/SCPS	Azomethine polyester	Tafel polarization and electrochemical impedance, FTIR, XRD, and AFM	An adsorptive layer was formed on the rebar under the simulated atmosphere.	45
36	Steel/SCPS	Sodium tungstate (Na₂WO₄) and hexamethylene tetramine	Electron Micro-Probe Analyzer, XPS, and Raman spectroscopy	Mixed type of inhibitor.	46
37	Steel/SCPS	Imidazoline quaternary ammonium salt	Polarization and EIS	Reduced current density.	47
38	Steel/SCPS	Diethanolamine	Weight loss measurements	This reduces the activity of the cathode and the anode by the formation of a consecutively protective film on the surface.	48
39	Carbon steel/ SCPS	1-Dihydroxyethylamino-3-dipropylamino-2-propanol (HPP)	Linear polarization resistance, EIS and CV, SEM, and XRD	HPP behaves as an anodic corrosion inhibitor.	49

S. no.	Metals used/ Medium	Inhibitor/Additive	Method	Findings	Ref.
40	Steel/SCPS	$C_4H_{13}N_3$	Weight loss measurement and polarization study	Prevents CR.	50
41	Steel/SCPS	Organic inhibitor	Polarization curves and EIS	The CR decreases with an increase in the concentration of the inhibitor.	51
42	Prestressed steel/SCPS	Ammonium molybdate, diethylenetriamine (DETA), propylene thiourea, and 1,4-butynediol	Structural studies	These prevent chloride ions from penetration and reduce the free rate of iron ions.	52
43	Carbon steel/ SCPS	Triethylentetramine, dimethylamine, N,N-dimethylethanolamine, ethanolamine, and 1, 6-hexanediamine	Electrochemical potentiodynamic polarization and EIS	The effectiveness of amine-based inhibitors is related to the pH value of the solution.	53
44	Mild steel/ SCPS	1,3-bis-dibutylaminopropan-2-ol	XPS	This effectively suppresses the anodic process of carbon steel corrosion.	54
45	Carbon steel/ SCPS	*Chamaerops humilis l.* leaves (MECHLL)	EIS and potentiodynamic polarization curves	Langmuir and Temkin adsorption isotherms.	55
46	Carbon steel/ SCPS	Amines and alkanolamines	Potentiodynamic polarization curves	Formation of a protective layer.	56

(Continued)

Table 9.1 (Continued)

S. no.	Metals used/ Medium	Inhibitor/Additive	Method	Findings	Ref.
47	Carbon steel/ SCPS	Phosphate-based inhibitor	–	Formation of a protective phosphate layer.	57
48	Steel/SCPS	1,3-bis-dialkylamino-2-propanol	Electrochemical techniques, XPS, and quantum chemical calculations	Langmuir and Temkin adsorption isotherms, mixed-type inhibitors.	58
49	Steel/SCPS	D-Sodium gluconate	Polarization curve	The inhibitor mainly acted as an anodic corrosion inhibitor.	59
50	Steel/SCPS	Imidazoline	–	Cationic corrosion inhibitor.	60
51	Steel/SCPS	Sodium D-gluconate	Conventional weight loss and polarization measurements techniques	Synergetic effect among the three agents in corrosion prevention.	61
52	Steel/SCPS	Sodium phosphate (Na_3PO_4), sodium nitrite ($NaNO_2$), and benzotriazole (BTA)	EIS	–	62
53	Steel/SCPS	Sodium nitrate	Weight loss measurements as well as potentiostatic technique	$NaNO_3$ reduces corrosion levels in basic environments.	63
54	C-steel, nickel, and zinc/SCPS	Aqueous extract of the leaves of henna (*Lawsonia*)	Polarization technique	Mixed inhibitor; Langmuir adsorption isotherm.	64

S. no.	Metals used/ Medium	Inhibitor/Additive	Method	Findings	Ref.
55	Mild steel/ SCPS	Ammonium carboxylate	Polarization, EIS, and XPS	The inhibitor has higher charge transfer resistance.	65
56	Steel/SCPS	Sodium molybdate	EIS and Mott–Schottky analysis technique	Langmuir adsorption isotherm.	66
57	Steel/SCPS	$NaNO_2$	Weight loss measurements, potentiodynamic polarization, and AC impedance measurements	Good corrosion inhibitor.	67
58	Steel/SCPS	*N*-lauroyl sarcosine sodium	Weight loss tests	Good inhibition efficiency.	68
59	Carbon steel/ SCPS	Benzotriazole (BTAH)	–	BTAH is a potentially attractive alternative to nitrites for inhibiting corrosion.	69
60	Mild steel/ SCPS	D-sodium gluconate	Electrochemical techniques and SEM	The IE of the inhibitor was 99%.	70
61	Steel/SCPS	Di-(2-ethylhexyl)-sebacate	Electrochemical techniques	Suppress the electrochemical reaction by forming a deposition film in the steel–concrete interface.	71
62	Steel/SCPS	D-sodium gluconate	Polarization and EIS	An adsorptive film is formed on the steel surface because the gluconate anions compete against the chloride ions in the adsorption.	72

(Continued)

Table 9.1 *(Continued)*

S. no.	Metals used/ Medium	Inhibitor/Additive	Method	Findings	Ref.
63	Steel/SCPS	Na_2PO_3	Electrochemical polarization curves and EIS	The electrochemical reaction can be suppressed by forming a deposition film on the steel.	73
64	Prestressed steel/SCPS	Nitrite	–	Embrittlement occurring in prestressed steel can be mainly due to the hydrogen generated during the corrosion process.	74
65	Steel/SCPS	Phosphate and nitrite	EIS, polarization curves, and OCP measurements, SEM, XRD, and FTIR	The results indicate that after 1 year of immersion with the two inhibitors, the CR decreased. Then, after 3 years of immersion, no influence of inhibitors on the CR was observed.	75
66	Carbon steel/ SCPS	5-Aminouracil	Polarization measurements	5-Aminouracil was adsorbed onto steel surfaces in accordance with the Langmuir adsorption isotherm.	76
67	Steel/SCPS	DETA/thiourea (TU)	Linear polarization, Tafel polarization, and EIS	DETA-TU and $NaNO_2$ had a significant synergistic effect on corrosion inhibition.	77
68	Carbon steel/ SCPS	Calcium nitrite	Potentiodynamic polarization studies and XPS	A decrease in pH was recorded after the inhibitor addition.	78

9.7 Case Study: Inhibition of Corrosion of Mild Steel in Simulated Concrete Pore Solution in the Presence of an Aqueous Extract of a Natural Product

The IE of *Hibiscus* in controlling corrosion of mild steel in SCPS has been evaluated by the weight loss method and electrochemical studies. The results are presented next.

9.7.1 Analysis of the Weight Loss Method

The rate of corrosion of mild steel immersed in SCPS prepared in well water in the absence and presence of *Hibiscus* leaf extract has been calculated by the weight loss method The calculated IEs and corrosion rates (CRs) of mild steel immersed in SCPS for a period of 1 day in the absence and presence of aqueous extract of *Hibiscus* leaf is given in Table 9.2.

Table 9.2 Inhibition efficiency (%) and corrosion rate (mdd) of mild steel immersed in a simulated concrete pore solution prepared in well water in the absence and presence of extract from *Hibiscus* leaf system (corrosion rate in well water = 49.72 mdd)

Hibiscus leaf extract + SCPS (mL)	Corrosion rate (mdd)	Inhibition efficiency (%)
0	9.45	81
2	23.86	52
4	18.39	63
6	13.92	72
8	8.452	83
10	2.9832	94

Various concentrations of the aqueous extract of *Hibiscus* leaves (2, 4, 6, 8, and 10 mL) were added to the SCPS system. The corrosion inhibition effectiveness slowly increased. The maximum IE was found to be 94%. Thus it was observed that when 10 mL of *Hibiscus* leaf extract was added to SCPS, the IE increased from 81% to 94%. First the IE decreased and then it increased. This may be due to the fact that in the initial state, the protective film formed was unstable, and then later, the protective film formed was more stable.

9.7.2 Analysis of Potentiodynamic Polarization Curves

The potentiodynamic polarization curves of mild steel immersed in well water in the absence and presence of an inhibitor system are shown in Fig. 9.1. When mild steel is immersed in well water, the linear polarization resistance (LPR) value is 23537.8 ohm cm^2. The corrosion current is 1.835×10^{-6} A/cm^2. When mild steel is immersed in SCPS prepared in well water, the corrosion potential shifts to the anodic side (–525 mV vs. saturated calomel electrode [SCE]). This indicates that there is passivation of metal by formation of a protective film on the anodic side of the metal surface. This is confirmed by an increase in the LPR value from 23537.8 to 26795.8 ohm cm^2. Further, the corrosion current decreases from 1.835×10^{-6} to 1.573×10^{-6} A/cm$^{2.}$

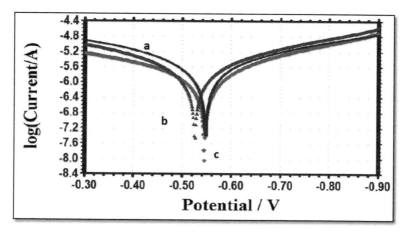

Figure 9.1 Polarization curve of mild steel immersed in various test solutions (Tafel plots): (a) well water, (b) SCPS prepared in well water, and (c) SCPS + *Hibiscus* leaf extract.

When 10 mL of aqueous extract of *Hibiscus* leaf is added, the corrosion potential shifts to –545 mV versus the SCE. When compared to the well water system, there is not much shift in the corrosion potential; this indicates that *Hibiscus* leaves combined with the SCPS system function as a mixed type of inhibitor, that is, both anodic and cathodic reactions are controlled to an equal extent. The LPR value increases to 38154.9 ohm cm^2; further the corrosion current decreases to 1.133×10^{-6} A/cm^2. Thus it is observed that when *Hibiscus* leaf extract is added, the corrosion resistance of mild steel in SCPS increases (Table 9.3).

Table 9.3 Corrosion parameters of mild steel immersed in well water and SCPS in the absence and presence of inhibitor *Hibiscus* leaf extract

System	E_{corr} (mV vs. SCE)	b_c (mV/ decade)	b_a (mV/ decade)	LPR (ohm cm^2)	I_{corr} (A/cm^2)
Well water	−547	207	191	23537.8	1.83×10^{-6}
SCPS	−525	191	197	26795.8	1.57×10^{-6}
SCPS + *Hibiscus* leaf extract	−545	180	222	38154.9	1.133×10^{-6}

9.7.3 Analysis of AC Impedance Spectra

When mild steel is immersed in well water, the charge transfer resistance (R_t) is 4080 ohm cm^2 and the double-layer capacitance (C_{dl}) is 12.5×10^{-10} F/cm^2. The impedance is equal to 3.805 log(z/ohm). When mild steel is immersed in SCPS, R_t increases from 4080 ohm cm^2 to 4680 ohm cm^2 and C_{dl} decreases from 12.5×10^{-10} F/cm^2 to 10.9×10^{-10} F/cm^2. The impedance value increases from 3.805 to 3.844 log(z/ohm). When mild steel is immersed in well water with SCPS and *Hibiscus* leaf extract, R_t increases to 6397 ohm cm^2 and C_{dl} decreases to 7.972×10^{-10} F/cm^2. The impedance value increases to 3.95 log(z/ohm). Thus the analysis of AC impedance spectra leads to the conclusion that in the presence of the SCPS + sitosterol system, the corrosion resistance of mild steel increases (Fig. 9.2, Fig. 9.3, and Table 9.4). This is due to the formation of a protective film formed metal surface. The protective film probably consists of an iron complex of the active principle present in *Hibiscus* leaves (Fe^{2+}-β-sitosterol complex).

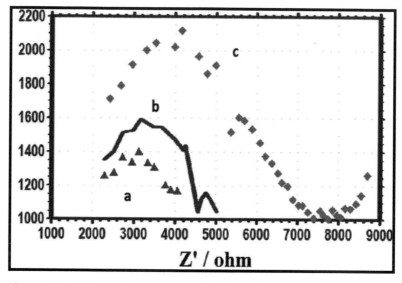

Figure 9.2 AC impedance spectra of mild steel immersed in various test solutions (Nyquist plots): (a) well water, (b) SCPS prepared in well water, and (c) SCPS + *Hibiscus* leaf extract.

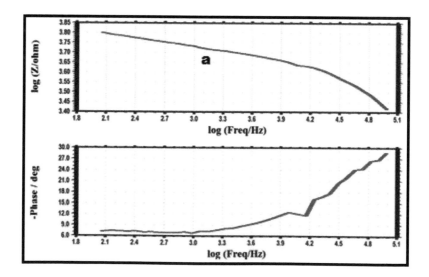

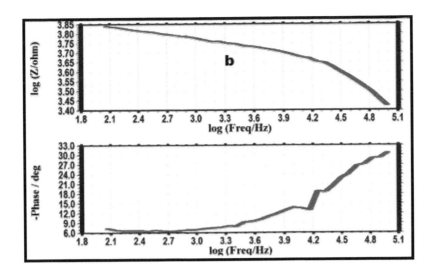

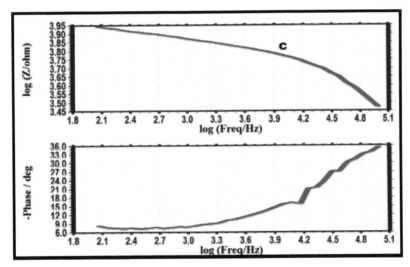

Figure 9.3 AC impedance spectra of mild steel immersed in various test solutions (Bode plots): (a) well water, (b) SCPS, and (c) SCPS + *Hibiscus* leaf extract.

Table 9.4 Corrosion parameters of mild steel immersed in simulated concrete pore solution in the absence and presence of inhibitor *Hibiscus* leaf extract, obtained from AC impedance spectra

| System | Nyquist plot | | Bode plot |
	R_t (ohm cm^2)	C_{dl} (F/cm^2)	Impedance value log(z/ohm)
Well water	4080	12.5×10^{-10}	3.805
SCPS	4680	10.9×10^{-10}	3.844
SCPS + *Hibiscus* leaf extract	6397	7.972×10^{-10}	3.95

9.7.4 AFM Studies

AFM images of mild steel immersed in SCPS and *Hibiscus* leaf extract are shown in Figs. 9.4 and 9.5. It is observed from Table 9.5, for polished mild steel, the root-mean-square (RMS) roughness (R_q) is 13.833 nm. The average roughness (R_a) is 11.485 nm. The maximum peak-to-valley height (R_y) is 72.147 nm. Analysis of Table 9.5 reveals that the R_q for mild steel immersed in SCPS is higher than that of polished mild steel. This indicates that due to corrosion, a thick nanofilm is formed on the metal surface. Similar is the case with R_a and R_y.

Table 9.5 AFM study values

Sample	RMS roughness (R_q) (nm)	Average roughness (R_a) (nm)	Maximum peak- to- valley height (R_y) (nm)
Polished metal	13.833	11.485	72.147
Metal immersed in well water	541.76	406.56	267.4
Mild steel immersed in SCPS	350.09	227.9	2113.5
Mild steel immersed in SCPS and 10 mL of *Hibiscus* leaf extract	321.75	62.3	1623.4

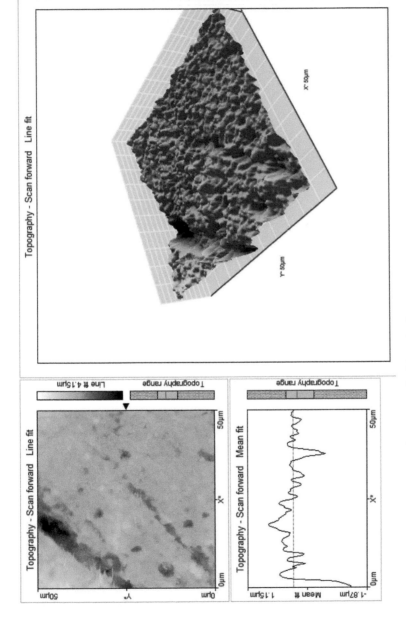

Figure 9.4 AFM images of mild steel in SCPS and *Hibiscus* leaf extract.

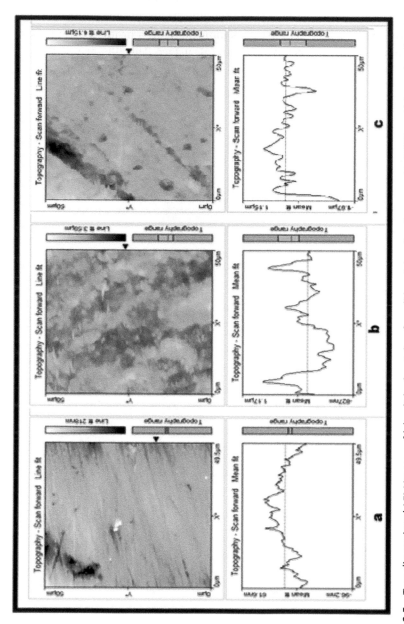

Figure 9.5 Two-dimensional AFM images of (a) polished metal surface, (b) metal surface immersed in SCPS, and (c) metal surface immersed in SCPS and *Hibiscus* leaf extract.

For a metal immersed in the inhibitor system, R_q is higher (321.75 nm) than that of polished metal but lower than that of metal immersed in a corrosive medium. Similar is the case with R_a (62.3 nm) and R_y (1623.4 nm). This indicates that a protective film of nanometer scale is formed on the metal surface. This film protects the metal from corrosion.

References

1. Andrade, C., Merino, P., Novoa, X. R., Perez, M. C., and Solar, L. (1995). *Mater. Sci. Forum*, **192–194**, p. 861.

2. Hansson, C. M. (1984). *Cem. Concr. Res.*, **14**, p. 574.

3. Ramirez, E., Gonzalez, J. A., and Bautista, A. (1996). *Cem. Concr. Res.*, **26**, p. 1525.

4. Hou, J., and Chung, D. D. L. (2000). *Corros. Sci.*, **42**, p. 1489.

5. Gonzalez, J. A., Miranda, J. M., Otero, E., and Feliu, S. (2007). *Corros. Sci.*, **49**, p. 436.

6. Nakayama, N. (2000). *Corros. Sci.*, **42**, p. 1897.

7. Nakayama, N., and Obuchi, A. (2003). *Corros. Sci.*, **45**, p. 2075.

8. Onzalez, J. A., Cobo, A., Gonzaalez, M. N., and Otero, E. (2000). *Mater. Corros.*, **51**, p. 97.

9. Blanco, D., Bautista, A., and Takenouti, H. (2006). *Cem. Concr. Compos.*, **28**, p. 212.

10. Lide, D. R. (1999). *CRC Handbook of Chemistry and Physics* (New York, NY: CRC Press).

11. Xu, J., Song, Y., Zhao, Y., et al. (2018). *Appl. Clay Sci.*, **163**, pp. 129–136.

12. Dong, S., Gao, Y., Guan, Z., et al. (2018). *Gaodeng Xuexiao Huaxue Xuebao/Chem. J. Chin. Univ.*, **39**(6), pp. 1260–1266.

13. AccessJi, T., Ma, F., Liu, D., et al. (2018). *Int. J. Electrochem. Sci.*, **13**(6), pp. 5440–5451.

14. Maliekkal, B. P., Kakkassery, J. T., and Palayoor, V. R. (2018). *Int. J. Ind. Chem.*, **9**(2), pp. 105–114.

15. Ben Mansour, H., Dhouibi, L., and Idrissi, H. (2018). *Constr. Build. Mater.*, **171**, pp. 250–260.

16. Zertoubi, M., Khoudali, S., and Azzi, M. (2018). *Port. Electrochim. Acta*, **36**(4), pp. 249–257.

17. Wang, Y.-Q., Kong, G., Che, C.-S., and Zhang, B. (2018). *Constr. Build. Mater.*, **162**, pp. 383–392.

18. Hassoune, M., Bezzar, A., Sail, L., and Ghomari, F. (2018). *J. Adhes. Sci. Technol.*, **32**(1), pp. 68–90.

19. Lin, B., Luo, Z., Liu, C., Liu, X., and Xu, Y. (2017). *Int. J. Electrochem. Sci.*, **12**(10), pp. 8892–8907.

20. Jiang, S. B., Jiang, L. H., Wang, Z. Y., et al. (2017). *Constr. Build. Mater.*, **150**, pp. 238–247.

21. Zhu, Y., Ma, Y., Yu, Q., Wei, J., and Hu, J. (2017). *Mater. Des.*, **119**, pp. 254–262.

22. Wang, Y., and Zuo, Y. (2017). *Corros. Sci.*, **118**, pp. 24–30.

23. Lin, B., Liu, C., Luo, Z., et al. (2017). *Int. J. Electrochem. Sci.*, **12**(3), pp. 2070–2087.

24. Liu, J. Z., Cai, J. S., Shi, L., et al. (2017). *Anti-Corros. Methods Mater.*, **64**(5), pp. 555–562.

25. Munot, H., Deshpande, P., and Modhera, C. (2017). *UPB Sci. Bull. Ser. B: Chem. Mater. Sci.*, **79**(2), pp. 191–206.

26. Guan, Z.-C., Gao, Y.-B., Wang, X., et al. (2017). *ECS Trans.*, **80**(10), pp. 663–672.

27. Wang, X.-P., Guan, Z.-C., Wang, X., Wang, H.-P., and Du, R.-G. (2017). *ECS Trans.*, **80**(10), pp. 655–661.

28. Bensabra, H., Franczak, A., Aaboubi, O., Azzouz, N., and Chopart, J.-P. (2017). *Metall. Mater. Trans. A*, **48**(1), pp. 412–424.

29. Lin, B., and Xu, Y. (2016). Jianzhu Cailiao Xuebao/*J. Build. Mater.*, **19**(6), pp. 1082–1087.

30. Verbruggen, H., Terryn, H., and De Graeve, I. (2016). *Constr. Build. Mater.*, **124**, pp. 887–896.

31. Wang, Y., Zuo, Y., Zhao, X., and Zha, S. (2016). *Appl. Surf. Sci.*, **379**, pp. 98–110.

32. Shubina, V., Gaillet, L., Chaussadent, T., Meylheuc, T., and Creus, J. (2016). *J. Cleaner Prod.*, **112**, pp. 666–671.

33. Liu, J. Z., Zhao, D., Cai, J. S., Shi, L., and Liu, J. P. (2016). *Int. J. Electrochem. Sci.*, **11**(2), pp. 1135–1151.

34. Bensabra, H., Azzouz, N., Aaboubi, O., and Chopart, J. P. (2016). *Metall. Res. Technol.*, **113**(1), p. 102.

35. Liu, J., Zhao, D., Cai, J., Shi, L., and Liu, J. (2016). *Int. J. Electrochem. Sci.*, **11**(10), pp. 8758–8778.

36. Yohai, L., Schreiner, W., Valcarce, M. B., and Vázquez, M. (2016). *J. Electrochem. Soc.*, **163**(13), pp. C729–C737.

37. Cabrini, M., Fontana, F., Lorenzi, S., Pastore, T., and Pellegrini, S. (2015). *J. Chem.*, **2015**, p. 521507.

38. 38.Al-Amoudi, O. S. B., Al-Sodani, K. A. A., and Maslehuddin, M. (2015). *NACE - International Corrosion Conference Series*, 2015-January.

39. Harish, E. L., Karthikeyan, S., and Sekar, S. K. (2015). *Int. J. ChemTech Res.*, **7**(4), pp. 2003–2006.

40. Devi, P. N., Rajendran, S., Sathiyabama, J., Jeyasundrai, J., and Umasankareswari, T. (2015). *Port. Electrochim. Acta*, **33**(3), pp. 195–200.

41. Li, J.-H., Zhao, B., Hu, J., et al. (2015). *Int. J. Electrochem. Sci.*, **10**(1), pp. 956–968.

42. Cao, F., Wei, J., Dong, J., and Ke, W. (2015). *Corros. Sci.*, **100**, pp. 365–376.

43. Lee, H.-S., Ryu, H.-S., Park, W.-J., and Ismail, M. A. (2015). *Materials*, **8**(1), pp. 251–269.

44. Jiang, J., Zhang, J., Lu, J., and Qu, W. (2015). *Jianzhu Cailiao Xuebao/J. Build. Mater.*, **18**(4), pp. 559–564 and 571.

45. Bhuvaneshwari, B., Selvaraj, A., Iyer, N. R., and Ravikumar, L. (2015). *Mater. Corros.*, **66**(4), pp. 387–395.

46. Gao, Y.-B., Hu, J., Zuo, J., et al. (2015). *J. Electrochem. Soc.*, **162**(10), pp. C555–C562.

47. Fei, F.-L., Hu, J., Wei, J.-X., Yu, Q.-J., and Chen, Z.-S. (2014). *Constr. Build. Mater.*, **70**, pp. 45–53.

48. Jiang, Y., Zheng, C. Y., and Liu, J. Z. (2014). *Appl. Mech. Mater.*, **507**, pp. 245–248.

49. Cai, J., Chen, C., Liu, J., and Liu, J. (2014). *Corros. Eng. Sci. Technol.*, **49**(1), pp. 66–72.

50. Sun, Q. L., Shi, L. D., and Liu, Z. R. (2014). *Appl. Mech. Mater.*, **556–562**, pp. 667–670.

51. Cai, J. S., Chen, C. C., Liu, J. Z., and Liu, J. P. (2014). *Proceedings of the 4th International Conference on the Durability of Concrete Structures*, ICDCS 2014, pp. 83–86.

52. Sun, Q. L., Liu, Z. R., Guo, Y. Z., and Liu, F. (2014). *Adv. Mater. Res.*, **941–944**, pp. 1390–1393.

53. Xu, C., Jin, W., and Zhang, S. (2014). *Jianzhu Cailiao Xuebao/J. Build. Mater.*, **17**(4), pp. 572–578.

54. Liu, J. P., Chen, C. C., Cai, J. S., Liu, J. Z., and Cui, G. (2013). *Mater. Corros.*, **64**(12), pp. 1075–1081.

55. Benmessaoud Left, D., Zertoubi, M., Khoudali, S., et al. (2013). *Int. J. Electrochem. Sci.*, **8**(10), pp. 11768–11781.

56. Zhang, S.-Y., Jin, W.-L., and Xu, C. (2013). *Zhejiang Daxue Xuebao (Gongxue Ban)/Journal of Zhejiang University (Engineering Science)*, **47**(3), pp. 449–455 + 487.

57. Nahali, H., Dhouibi, L., and Idrissi, H. (2013). *Ann. Chim. Sci. Mat.*, **38**(1–2), pp. 49–58.

58. Cai, J., Chen, C., Zhou, W., and Liu, J. (2012). *Int. J. Electrochem. Sci.*, **7**(11), pp. 10894–10908.

59. Wang, X.-P., Du, R.-G., Zhu, Y.-F., et al. (2012). *ECS Trans.*, **50**(50), pp. 43–51.

60. Yu, Q.-J., Fei, F.-L., Wei, J.-X., Hu, J., and Ai, Z.-Y. (2012). *Huanan Ligong Daxue Xuebao/Journal of South China University of Technology (Natural Science)*, **40**(10), pp. 134–141.

61. Yang, R.-J., Guo, Y., Tang, F.-M., et al. (2012). *Wuli Huaxue Xuebao/Acta Physico - Chimica Sinica*,

62. Shi, J.-J., and Sun, W. (2012). *Int. J. Miner. Metall. Mater.*, **19**(1), pp. 38–47.

63. Saura, P., Zornoza, E., Andrade, C., and Garcés, P. (2011). Steel corrosion-inhibiting effect of sodium nitrate in simulated concrete pore solutions, *Corrosion*, **67**(7), pp. 075005-1–075005-15.

64. Shi, J.-J., and Sun, W. (2010). *Gongneng Cailiao/J. Funct. Mater.*, **41**(12), pp. 2147–2150.

65. Miao, C., Zhou, W., and Chen, C. (2010). *Dongnan Daxue Xuebao (Ziran Kexue Ban)/Journal of Southeast University (Natural Science Edition)*, **40**(Suppl. 2), p. 187.

66. Zhou, X., Yang, H.-Y., and Wang, F.-H. (2010). *Corros. Sci. Prot. Technol.*, **22**(4), pp. 343–347.

67. Qiao, B., Du, R., Chen, W., Zhu, Y., and Lin, C. (2010). *Jinshu Xuebao/Acta Metall. Sinica*, **46**(2), pp. 245–250.

68. Qiao, B., Du, R.-G., and Lin, C.-J. (2009). *Gongneng Cailiao/J. Funct. Mater.*, **40**(9), pp. 146–149.

69. Mennucci, M. M., Banczek, E. P., Rodrigues, P. R. P., and Costa, I. (2009). *Cem. Concr. Compos.*, **31**(6), pp. 418–424.

70. Chen, W., Wu, Q., Du, R.-G., Lin, C.-J., and Sun, L. (2009). *Gongneng Cailiao/J. Funct. Mater.*, **40**(4), pp. 611–613.

71. Chen, Y., Wang, W., and Yang, J. (2007). *Beijing Keji Daxue Xuebao/ Journal of University of Science and Technology Beijing*, **29**(Suppl. 2), pp. 21–24.

72. Li, J.-H., Zhao, B., Du, R.-G., and Lin, C.-J. (2007). *Gongneng Cailiao/J. Funct. Mater.*, **38**(3), pp. 509–511.

73. Zhang, D.-Q., Zhang, W.-Y., Wang, W., and Zhou, G.-D. (2007). *Corros. Prot.*, **28**(2), p. 55.

74. Alonso, C., Fullea, J., and Andrade, C. (2003). *J. Corros. Sci. Eng.*, **6**.

75. Dhouibi, L., Triki, E., Salta, M., Rodrigues, P., and Raharinaivo, A. (2003). *Mater. Struct./Materiaux et Constructions*, **36**(262), pp. 530–540.

76. Nakayama, N. (2002). *Electrochemistry*, **70**(5), pp. 322–328.

77. Wang, S.-X., Lin, W.-W., Zhang, J.-Q., and Fang, Z.-K. (2000). *J. Chin. Soc. Corros. Prot.*, **20**(1), pp. 20–21.

78. Tullmin, M., Mammoliti, L., Sohdi, R., Hansson, C. M., and Hope, B. B. (1995). *Cem. Concr. Aggregates*, **17**(2), pp. 134–144.

Chapter 10

Corrosion Inhibitors for Metals and Alloys in a Neutral Medium

Seawater, well water, and water containing low chloride ions can be used in cooling water systems. Various types of inhibitors can be used to control the corrosion of pipes carrying water. Methods employed and surface analysis techniques are discussed in this chapter.

10.1 Inhibitors

Various types of inhibitors have been used to control corrosion of metals in a neutral medium. In this connection, mention may be made of organic polymers [1], melamine derivatives [2], benzotriazole [3], drugs [4], benzimidazole [3], polyphenols [10], tolyltriazole [14], amino acids [15], phosphonic acids [18, 23, 24] and polyvinyl pyrrolidone (PVP) [25, 26].

The various aspects are summarized in Table 10.1.

Titanic Corrosion
Susai Rajendran and Gurmeet Singh
Copyright © 2020 Jenny Stanford Publishing Pte. Ltd.
ISBN 978-981-4800-95-2 (Paperback), 978-1-003-00449-3 (eBook)
www.jennystanford.com

Table 10.1 Corrosion inhibitors for metals/alloys in a neutral medium

S. no.	Metal/Medium	Inhibitor/Additives	Methods	Findings	Ref
1	Mild steel/NaCl and $(NH_4)_2SO_4$	Organic polymer matrix	Fourier transform infrared spectroscopy (FTIR), X-ray diffraction (XRD), scanning electron microscopy (SEM), electrochemical impedance spectroscopy (EIS)	According to Bustos-Terrones et al. (2018), the results of the simulation suggest that base polyaniline (PANIB) has less affinity for fluconazole (F) than polyaniline salt (PANIS), which favors the dissolution when the electrolyte penetrates the pores of the coating.	1
2	Mild steel /(1M HCl)	Melamine (triazine) and glyoxal, namely 2,2-bis(4,6-diamino-1,3,5-triazin-2-ylamino) acetaldehyde (ME-1) and (N2,N2'E,N2,N2'E)–N2,N2'-(ethane-1,2-diylidene)-bis-(1,3,5-triazine-2,4,6-triamine) (ME-2)	Electrochemical analyses, surface characterization methods, energy of highest occupied molecular orbital (E_{HOMO}) and energy of lowest unoccupied molecular orbital (E_{LUMO}), density functional theory (DFT) studies	The inhibition efficiency of ME-1 and ME-2 increases with an increase in their concentrations and maximum values of 91.47% and 94.88%, Langmuir adsorption isotherm, smoothness in the surface morphologies.	2
3	Zn-Mg-coated steel corrosion in chloride electrolyte	Benzotriazole	Electrochemical and analytical measurements	Corrosion reactions by forming a complex between Zn^{2+} and the anionic form of benzotriazole, pH increases, corrosion product layer composed of zincite (ZnO), which precipitates.	3

S. no.	Metal/Medium	Inhibitor/Additives	Methods	Findings	Ref
4	Corrosion of copper in acid media, neutral medium	Eco-friendly drug clozapine	Weight loss, EIS, and potentiodynamic polarization studies	Good corrosion inhibition efficiency, Langmuir adsorption isotherm.	4
5	Metal/in acid, alkaline, and near-neutral media	Benzimidazole	The mechanisms of corrosion protection for each inhibition system have also been highlighted via experimental- and theoretical-based approaches in acid, alkaline, and near-neutral media.	Good corrosion inhibition efficiency, adsorption isotherm, sp2 electron pair.	5
6	Corrosion inhibition of copper in aqueous chloride	Phosphorylated chitin (PCT)	Investigated by gravimetric analysis, EIS, polarization, X-ray photoelectron spectroscopy (XPS), SEM, energy-dispersive X-ray spectroscopy (EDX)	Maximum efficiency of 92%, chemical composition of the film formed on the metal surface.	6

(*Continued*)

Table 10.1 (*Continued*)

S. no.	Metal/Medium	Inhibitor/Additives	Methods	Findings	Ref
7	1,3-Diphenylthiourea-doped sol-gel coating on aluminum	3-Aminopropyltrimethoxysilane (APTMS)	FTIR, SEM, EDX, and thermogravimetric analysis (TGA)	The inhibitor-doped sol-gel coating increased the hydrophobic nature of the coating, adherent layer over metallic surface, improved corrosion resistance.	7
8	Steel/Neutral aqueous salt and acid media	Synergistic compositions	Method of isomolar series, synergistic effect	Increase their efficiency, result of the improvement of the microstructure of the phase layer. Isomolar series, of synergistic effects.	8
9	Carbon steel AISI 1020 in a neutral medium	Coconut (*Cocos nucifera*)	Electrochemical analysis (linear polarization resistance)	Cathodic corrosion inhibitor: The inhibition efficiency was increased by 90%, suggesting the coconut tannin may react with the metal surface protecting it. Anticorrosive agent.	9
10	Natural plant extracts or biodegradable material	Polyphenols	Weight loss tests	Increase their efficiency, neutral medium, whose value was 74%. The efficiency of these inhibitors was more significant in 1 mol·L^{-1} hydrochloric acid media, with efficiencies above 93%.	10

S. no.	Metal/Medium	Inhibitor/Additives	Methods	Findings	Ref
11	Mild steel in neutral and acid media	Chitin or poly(N-acetyl-1,4-β-d-glucopyranosamine)	Weight loss, potentiodynamic polarization, electrochemical impedance techniques, SEM, XPS	Result of the improvement of the microstructure of the phase layer, isomolar series, synergistic effects. Inhibition efficiency of 94%.	11
12	Copper and carbon steel/neutral (NaCl)/acidic (HCl)	Petroleum and synthetic (poly-α-olefins) oils	Kinetic parameters (XPS)	Results of the high porosity of the used oil films regardless of their kind.	12
13	Aluminum alloy metal in neutral medium (NaCl)	Synthetic (poly-α-olefins) oils	Weight loss, potentiodynamic, and cyclic polarization techniques. SEM and EDX	The result shows that intermetallic phases may directly influence the corrosion behavior of the aluminum alloys.	13
14	Steel in chloride media	Azole inhibitors and metal salts, tolyltriazole	Attenuated total reflection (ATR)-FTIR analysis, atomic force microscopy (AFM), electrochemical techniques	The most effective corrosion inhibitor; adsorption of tolyltriazole on the steel surface, reduces surface roughness.	14

(*Continued*)

Table 10.1 *(Continued)*

S. no.	Metal/Medium	Inhibitor/Additives	Methods	Findings	Ref
15	Steel in neutral water medium	Fatty acid and amino acid	Weight loss method, SEM	Inhibition efficiency increased to adsorption equilibrium in the metal surface. Using SEM to observe the microcorrosion morphology.	15
16	Copper, sulfureus in neutral medium	*Azadirachta indica* (neem) leaf extract	Weight loss, EIS, polarization studies, SEM	Effective in inhibiting corrosion.	16
17	Mild steel in chloride media	Ethanol extract of propolis	Electrochemical and electrochemical probe beam deflection techniques	Effective metal corrosion inhibitors	17
18	Carbon steel in aqueous solution	Propyl phosphonic acid	Investigated by gravimetric studies, potentiodynamic polarization, EIS	Corrosion inhibition is presented, protective film formed on the iron substrate, mixed type (anodic and cathodic).	18
19	Carbon steel in neutral medium	*Citrus medica*	Mass loss measurements, FTIR and UV studies	Inhibition efficiency is increased with increase of inhibitor concentration.	19

S. no.	Metal/Medium	Inhibitor/Additives	Methods	Findings	Ref
20	Iron in acidic and alkaline media	Aminic nitrogen-bearing polydentate Schiff base	Polarization study	DFT study gave further insight into the mechanism of inhibition action of polydentate Schiff base compounds (PSCs). Madkour and Elroby [20] provided an approach to evaluating the theoretical inhibition performance for homologous inhibitors. The inhibition efficiency of PSCs increased with an increase in E_{HOMO} and a decrease in E_{LUMO}-E_{HOMO}.	20
21	Steel in neutral aqueous salt solutions	Aqueous salt solutions	Electrochemical studies	Anodic reaction (ionization of metal) is controlled predominantly.	21
22	Mild steel in acidic HCl and neutral media	*Vitex negundo* leaf extract	Weight loss method, polarization study	Inhibition efficiency increased with increasing concentration of extract. A protective film is formed on the metal surface.	22
23	Steel and high salt content	Phosphonate inhibitors in waters with a high salt content	–	Mild steel can be successfully protected from corrosion by the magnesium complex of phosphonic acids in combination with small amounts of nitrogen-containing corrosion inhibitors (catamine AB, IFKhAN-92) in waters with a high salt content at low content of H_2S and in its absence.	23

(Continued)

Table 10.1 (*Continued*)

S. no.	Metal/Medium	Inhibitor/Additives	Methods	Findings	Ref
24	Carbon steel	Propyl phosphonic acid	Gravimetric method, impedance studies, XPS, FTIR, AFM, and SEM	Adsorbed protective film on the carbon steel surface.	24
25	Carbon steel alkaline solutions containing NaCl, inhibitor	Polyvinyl pyrrolidone (PVP)	Weight loss method, surface studies, EIS	Tafel results indicated that PVP is a mixed-type inhibitor. Increased inhibition efficiency.	25
26	Carbon steel in neutral solutions containing NaCl	PVP as a green corrosion inhibitor	EIS and potentiodynamic polarization	The adsorption of PVP led to the formation of a protective film on the metal/solution interface and an increase in the thickness of the electronic double layer, as revealed by EIS results.	26
27	Aluminum alloy	Cerium nitrate	EIS measurements	The Ce-P-Pr inhibitor combination (Ce^{3+}, PO_4^{3-}, Pr^{3+}) has shown the best long-term corrosion protection properties at low doping levels.	27

S. no.	Metal/Medium	Inhibitor/Additives	Methods	Findings	Ref
28	Mild steel neutral aqueous solution	Sodium molybdate and dicarboxylates	Potentiodynamic polarization, EIS, scanning vibrating electrode technique (SVET), AFM also used, surface analysis (SEM, XPS)	Protective film obtained, inhibitors at the coating damages, the protective layer formed in their presence is stable.	28
29	Al/NaOH system	p-Phenol derivatives	Electrochemical, thermodynamical	The results were amazing and revealed a complicated protonation/deprotonation cycle for inhibitor species inside electrical double layer.	29
30	Mild steel corrosion in neutral aqueous solution	Dicarboxylates, benzotriazole	Polarization curves and EIS	A synergistic effect, inhibitors show very high efficiency.	30
31	Iron nitric acid and sodium hydroxide	Schiff bases	Electrochemical techniques	The Tafel plot is proportional to its concentration. The adsorption process was found to obey Temkin adsorption isotherm.	31

(Continued)

Table 10.1 (*Continued*)

S. no.	Metal/Medium	Inhibitor/Additives	Methods	Findings	Ref
32	Iron and aqueous solution	Sodium nitrite inhibitor	Polarization curves and EIS, XRD, SEM, XRD patterns	Results obtained from electrochemical tests suggested that the inhibition effect and corrosion protection of sodium nitrite inhibitor in near-neutral aqueous solutions increased as the grain size decreased from submicrocrystalline to nanocrystalline.	32
33	Carbon steel	(H_2O, H_3O^+, and HCO^{3-})	1-R1-2-undecyl-imidazoline inhibitors: R1: CH2COOH (A), CH2CH2OH (B), CH2CH2NH2 (C)	Based on the above analysis, we found that theoretically the corrosion inhibition performance of the four imidazoline inhibitors decreased as follows: A>B>C>D, which agrees with previous experimental results.	33
34	Carbon steel in nearly neutral aqueous	*N,N*-bis(phosphonomethyl) glycine (BPMG)	Potentiodynamic polarization studies, XPS, SEM, FTIR	The calculated results showed that all the membranes formed by corrosion inhibitor molecules can effectively control corrosion of carbon steel.	34

S. no.	Metal/Medium	Inhibitor/Additives	Methods	Findings	Ref
35	Stainless steel in near-neutral chloride	Polyethyleneimines (PEI)	Linear polarization measurements, cyclic polarization measurements	Polymer adsorption was obtained by XPS. PEI molecules most probably physisorb on the metal surface, since no peak shift was observed that would indicate a C-N-metal connection.	35
36	Stainless steel in a 3% NaCl solution	PEIs	Linear polarization measurements and immersion	Polymer adsorption was obtained by XPS. PEI molecules most probably physisorb on the metal surface, since no peak shift was observed that would indicate a C-N-metal connection.	36
37	Vanadium in aqueous solutions	Amino acids	Polarization techniques and EIS	Corrosion inhibition efficiency, especially in acid solutions, was investigated. The adsorption free energy for valine on V in acidic solutions was found to be −9.4 kJ/mol, which reveals strong physical adsorption of the amino acid molecules on the vanadium surface.	37

(*Continued*)

Table 10.1 *(Continued)*

S. no.	Metal/Medium	Inhibitor/Additives	Methods	Findings	Ref
38	Copper	Hydroxamic acids [$CH_3(CH_2)16CONHOH$]	Different pH of the subphase Brewster angle microscope (BAM), AFM	The octadecanoyl hydroxamic acid (C 18N) results in a better and more stable monolayer with cations in the subphase. Langmuir layers were deposited.	38
39	Iron steel in a neutral solution	Triazole thiol and amino triazole	Electrochemical techniques	The inhibitor increases the polarization resistance, adsorption of the inhibitor through the N atom.	39
40	Magnesium/Sodium decanoate	Alkyl carboxylate	XRD, electrochemical measurements	The presence of structural water is confirmed by thermal analyses.	40
41	Brass in a neutral medium	Benzotriazole, N-[1-(benzotriazol-1-yl)benzyl] aniline (BTBA)	Weight loss measurements, polarization, EIS, cyclic voltammetry, and current transient techniques	Good inhibition efficiency in sodium chloride solution. Polarization studies showed that the benzotriazole derivatives are mainly anodic inhibitors for brass in a sodium chloride solution.	41
42	Boiler steel alkaline or acidic conditions neutral and alkaline media	Thiourea as an inhibitor	Electrochemical polarization technique, SEM and XRD studies	Results show that the parent metal experiences a higher corrosion rate than weld metal, and thiourea as an inhibitor is found unsuitable in a NaOH medium. Measure the corrosion current.	42

S. no.	Metal/Medium	Inhibitor/Additives	Methods	Findings	Ref
43	Muntz metal in acidic and neutral solutions	Triazole derivatives	Potentiodynamic polarization methods, UV spectral analysis, XRD, and SEM	The studied inhibitor molecules function as good corrosion inhibitors by being chemisorbed on metal surface. Among these compounds, 4-amino-5-mercapto-3-propyl-1,2,4-triazole (AMPT) is found to be more efficient.	43
44	Mustard oil in neutral and alkaline media	Alkyl- and arylbenzoimidazole	–	Mixtures of diethylammonium salts of sulfated mustard oil with benzoimidazole derivatives were found to be promising combined inhibitors of metal corrosion.	44
45	Steel in alkaline solutions	Ca(OH)2 + 0.1 M NaCl solution	Trisodium phosphate, calcium nitrite	The passive oxide film, which is formed on the metal surface in the presence of this compound, is more compact and protective in comparison to the oxide layer formed in the media without the inhibitor.	45
46	C-steel, nickel, and zinc in acidic, neutral, and alkaline solutions	Aqueous extract of the leaves of henna (*Lawsonia*)	Polarization technique	The inhibition efficiency increases as the added concentration of the extract is increased.	46

(Continued)

Table 10.1 (*Continued*)

S. no.	Metal/Medium	Inhibitor/Additives	Methods	Findings	Ref
47	Mild steel in neutral chloride media	Phosphonic acid	EIS, weight loss method	By increasing the concentration of citrate, the corrosion inhibition efficiency decreases.	47
48	Aluminum alloy in chloride mediums	Polyaniline (PANI)	EIS	The results indicated that corrosion protection of the bare aluminum alloy surface resulting from a defect of the PANI coating is in line with an increase of the thickness of the oxide layer protecting the aluminum alloy surface. The efficiency of corrosion protection of mild steel and aluminum alloy by PANI coatings was compared.	48
49	Carbon steel corrosion in acid media	1-(2-Ethylamino)-2-methylimidazoline	Potentiodynamic polarization, EIS	Good corrosion inhibitor at different concentrations (two nitrogen atoms) and the plane geometry of the heterocyclic ring, thus promoting coordination with the metal surface.	49
50	Ferrous and nonferrous materials in acidic and neutral media	Polymeric materials	Chemical and electrochemical techniques	Polymers that act both as antiscalants and as inhibitors.	50

10.2 Case Study: Corrosion Behavior of L 80 Alloy Immersed in Seawater in the Presence of Sodium Potassium Tartrate

Corrosion resistance of L 80 alloy immersed in various test solutions was measured by polarization studies.

10.2.1 Polarization Study

Polarization studies were carried out in a CHI Electrochemical workstation/analyzer, model 660A. It was provided with automatic infrared (IR) compensation facility. A three-electrode cell assembly was used.

The working electrode was L 80 alloy. A saturated calomel electrode (SCE) was the reference electrode. Platinum was the counterelectrode. A time interval of 5 to 10 min was given for the system to attain a steady-state open-circuit potential. The working electrode and the platinum electrode were immersed in seawater in the absence and presence of an inhibitor. From polarization studies, corrosion parameters such as corrosion potential (E_{corr}), corrosion current (I_{corr}), Tafel slopes anodic = b_a and cathodic = b_c, and linear polarization resistance (LPR) value were obtained. The scan rate (V/S) was 0.01. Hold time at (E_{fcs}) was 0 and quiet time (s) was 2.

10.2.2 Seawater

The composition of seawater used in this study is given in Table 10.2. Seawater was collected from the Bay of Bengal, located at Kanampadi, East Coast Road, Chennai, India.

Table 10.2 Composition of seawater

Physical examination	Acceptable limit	Permissible limit	Sample value
1. Color			Colourless & Clear
2. Odor	Unobjectionable		Unobjectionable
3. Turbidity NT units	1	5	0.2

4. Total dissolved solids mg/L1	500	2000	29400
5. Electrical conductivity micro mho/cm	–	–	42000
Chemical examination			
6. pH	6.5–8.5	6.5–8.5	7.46
7. pH alkalinity as $CaCO_3$	–	0	0
8. Total alkalinity as $CaCO_3$	200	600	140
9. Total hardness as $CaCO_3$	200	600	4000
10. Calcium as Ca	75	200	1200
11. Magnesium as Mg	30	100	240
12. Iron as Fe	0.1	1	0
13. Magnesium as Mg	0.1	0.3	NT
14. Free ammonia as NH_3	0.5	0.5	0.48
15. Nitrite as NO_2	0.5	0.5	0.104
16. Nitrate as NO_3	45	45	25
17. Chloride as Cl	250	1000	15000
18. Fluoride as F	1	1.5	1.8
19. Sulfate as SO_4	200	400	1170
20. Phosphate as PO_4	0.5	0.5	1.47
21. Tids test 4 h as O_2	–	–	NT

10.2.3   L 80 Alloy

L80 alloy was manufactured to API specification 5CT. This is a controlled yield strength material with a hardness testing requirement. L 80 alloy is usually used in wells with sour (hydrogen sulfide) environments. The chemical composition of L 80 alloy is given in Table 10.3.

Table 10.3   Chemical composition of L 80 alloy

	C	Mn	Ni	Cu	P	S	Si
Max.	0.430	1.900	0.250	0.350	0.030	0.030	0.450

The remaining wt% was iron.

10.2.4 Inhibitors Used

A pure sample of sodium potassium tartrate (SPT) and zinc sulfate were used as corrosion inhibitors.

10.2.5 Results

The polarization curves are shown in Figs. 10.1–10.4. The corrosion parameters are given in Table 10.4.

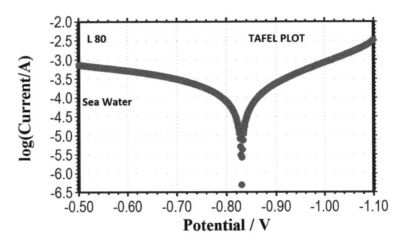

Figure 10.1 Polarization curve of L 80 alloy immersed in seawater.

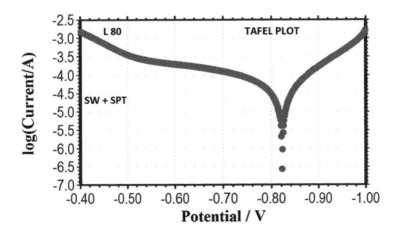

Figure 10.2 Polarization curve of L 80 alloy immersed in seawater + SPT.

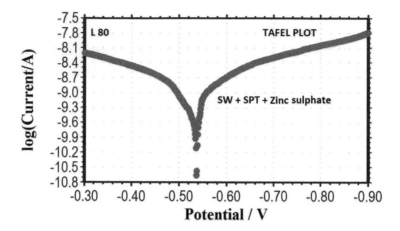

Figure 10.3 Polarization curve of L 80 alloy immersed in seawater + SPT + zinc sulfate.

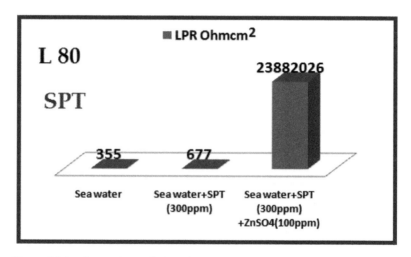

Figure 10.4 Comparison of LPR values.

It is seen from Table 10.4 that when SPT is added to seawater, the corrosion resistance of L 80 alloy increases. This is revealed by the fact that the LPR increases and I_{corr} decreases (Figs. 10.1–10.4). When $ZnSO_4 \cdot 7H_2O$ is added to the above system, the LPR increases tremendously (Fig. 10.4) and I_{corr} decreases a lot. This indicates that in the presence of SPT + $ZnSO_4 \cdot 7H_2O$ system, the corrosion resistance of L 80 alloy in seawater increases to a great extent. A protective film is formed on the metal surface. This prevents the transfer of electrons from the anodic site to the cathodic site. Hence, LPR increases, which results in a decrease in the corrosion current value.

Table 10.4 Corrosion parameters of L 80 alloy immersed in seawater in the absence and presence of SPT and zinc sulfate ($ZnSO_4 \cdot 7H_2O$) obtained by polarization studies.

System	E_{corr} (mV$_{SCE}$)	b_c (mV/ decade)	b_a (mV/ decade)	LPR (ohm·cm²)	I_{corr} (A/cm²)
Seawater	–830	161	221	355	1.138×10^{-4}
Seawater + SPT (300 ppm)	–822	119	212	677	4.908×10^{-5}
Seawater + SPT (300 ppm) + ZnSO$_4$ (100 ppm)	–537	209	252	23882026	2.079×10^{-9}

10.3 Conclusion

The corrosion resistance of L 80 alloy in seawater in the presence of SPT has been evaluated by polarization studies. It is observed that in the presence of SPT, the corrosion resistance of L 80 alloy in seawater increases. When zinc sulfate is added, the corrosion protection further increases. The decreasing order of corrosion resistance is as follows:

Seawater + Zinc sulfate + SPT > Seawater + Zinc sulfate > Seawater

References

1. Bustos-Terrones, V., Serratos, I. N., Castañeda-Villa, N., et al. (2018). Functionalized coatings based on organic polymer matrix against the process of corrosion of mild steel in neutral medium, *Prog. Org. Coat.*, **119**, pp. 221–229.

2. Verma, C., Haque, J., Ebenso, E. E., and Quraishi, M. A. (2018). Melamine derivatives as effective corrosion inhibitors for mild steel in acidic solution: chemical, electrochemical, surface and DFT studies, *Results Phys.*, **9**, pp. 100–112.

3. Rodriguez, J., Mouanga, M., Roobroeck, A., et al. (2018). Study of the inhibition ability of benzotriazole on the Zn-Mg coated steel corrosion in chloride electrolyte, *Corros. Sci.*, **132**, pp. 56–67.

4. Kumar, P. E., Govindaraju, M., and Sivakumar, V. (2018). Experimental and theoretical studies on corrosion inhibition performance of an environmentally friendly drug on the corrosion of copper in acid media, *Anti-Corros. Methods Mater.*, **65**(1), pp. 19–33.

5. Obot, I. B., and Edouk, U. M. (2017). Benzimidazole: small planar molecule with diverse anti-corrosion potentials, *J. Mol. Liq.*, **246**, pp. 66–90.

6. Vimal Kumar, K., Appa Rao, B. V., and Hebalkar, N. Y. (2017). Phosphorylated chitin as a chemically modified polymer for ecofriendly corrosion inhibition of copper in aqueous chloride environment, *Res. Chem. Intermed.*, **43**(10), pp. 5811–5828.

7. Vignesh, R. B., Balaji, J., and Sethuraman, M. G. (2017). Surface modification, characterization and corrosion protection of 1,3-diphenylthiourea doped sol-gel coating on aluminium, *Prog. Org. Coat.*, **111**, pp. 112–123.

8. Ledovs'kykh, V. M., Vyshnevs'ka, Y. P., Brazhnyk, I. V., and Levchenko, S. V. (2017). Development and optimization of synergistic compositions for the corrosion protection of steel in neutral and acid media, *Mater. Sci.*, pp. 1–9.

9. Ledovs'kykh, V. M., Vyshnevs'ka, Y. P., Brazhnyk, I. V., and Levchenko, S. V. (2017). Development and optimization of synergistic compositions for the corrosion protection of steel in neutral and acid media, *Mater. Sci.*, **52**(5), pp. 634–642.

10. da Rocha, J. C., and Gomes, J. A. C. P. (2017). Natural corrosion inhibitors-proposal to obtain ecological products of low cost from industrial waste [Inibidores de corrosão naturais- Proposta de obtenção de produtos ecologicos de baixo custo a partir de resíduos industriais], *Rev. Materia*, **22**, p. 11927.

11. Vimal, K., and Appa, R. B. V. (2017). Chemically modified biopolymer as an eco-friendly corrosion inhibitor for mild steel in a neutral chloride environment, *New J. Chem.*, **41**(14), pp. 6278–6289.

12. Vigdorovich, V. I., Tsygankova, L. E., Shel', N. V., et al.(2016). Kinetics and mechanism of electrode reactions in corrosion of some metals

covered with oil films in acid and neutral chloride environments, *Prot. Met. Phys. Chem.*, **52**(7), 1157–1165.

13. Loto, R. T., and Adeleke, A. (2016). Corrosion of aluminum alloy metal matrix composites in neutral chloride solutions, *J. Fail. Anal. Prev.*, **16**(5), pp. 874–885.

14. Finšgar, M., Petovar, B., Xhanari, K., and Maver, U. (2016). The corrosion inhibition of certain azoles on steel in chloride media: electrochemistry and surface analysis, *Corros. Sci.*, **111**, pp. 370–381.

15. Cheng, P., Huang, X.-Q., and Fu, C.-Y. (2016). Synthesis of neutral medium inhibitor and study of its inhibition efficiency, *MATEC Web of Conferences*, **67**, p. 04003.

16. Swaroop, B. S., Victoria, S. N., and Manivannan, R. (2016). Azadirachta indica leaves extract as inhibitor for microbial corrosion of copper by Arthrobacter sulfureus in neutral pH conditions: a remedy to blue green water problem, *J. Taiwan Inst. Chem. Eng.*, **64**, pp. 269–278.

17. Dolabella, L. M. P., Oliveira, J. G., Lins, V., Matencio, T., and Vasconcelos, W. L. (2016). Ethanol extract of propolis as a protective coating for mild steel in chloride media, *J. Coat. Technol. Res.*, **13**(3), pp. 543–555.

18. Prabakaran, M., Ramesh, S., Periasamy, V., and Sreedhar, B. (2015). The corrosion inhibition performance of pectin with propyl phosphonic acid and Zn2+ for corrosion control of carbon steel in aqueous solution, *Res. Chem. Intermed.*, **41**(7), pp. 4649–4671.

19. Yamuna, J., and Anthony, N. (2015). Corrosion protection of carbon steel in neutral medium using Citrus medica [CM] leaf as an inhibitor, *Int. J. ChemTech Res.*, **8**(7), pp. 318–325.

20. Madkour, L. H., and Elroby, S. K. (2014). Aminic nitrogen- bearing polydentate Schiff base compounds as corrosion inhibitors for iron in acidic and alkaline media: a combined experimental and DFT studies, *J. Corros. Sci. Eng.*, **17**.

21. Ledovs'Kykh, V. M., Levchenko, S. V., and Tulainov, S. M. (2014). Synergistic extrema of the mixtures of corrosion inhibitors for metals in aqueous salt solutions, *Mater. Sci.*, **49**(6), pp. 827–832.

22. Vimala, J. R., and Raja, S. (2013). Vitex negundo leaf extract as eco friendly inhibitor on the corrosion of mild steel in acidic and neutral media, *Res. J. Chem. Environ.*, **17**(12), pp. 84–91.

23. Redkina, G. V., Kuznetsov, Yu. I., and Frolova, L. V. (2013). Protection of steel by phosphonate inhibitors in waters with a high salt content, *EUROCORR 2013: European Corrosion Congress.*

24. Prabakaran, M., Venkatesh, M., Ramesh, S., and Periasamy, V. (2013). Corrosion inhibition behavior of propyl phosphonic acid-Zn2+system for carbon steel in aqueous solution, *Appl. Surf. Sci.*, **276**, pp. 592–603.

25. Juhaiman, L. A. A., Mustafa, A. A., and Mekhamer, W. K.　(2013). Polyvinyl pyrrolidone as a green corrosion inhibitor for carbon steel in alkaline solutions containing NaCl, *Anti-Corros. Methods Mater.*, **60**(1), pp. 28–36.

26. Juhaiman, L. A. A., Mustafa, A. A., and Mekhamer, W. K. (2012). Polyvinyl pyrrolidone as a green corrosion inhibitor of carbon steel in neutral solutions containing NaCl: electrochemical and thermodynamic study, *Int. J. Electrochem. Sci.*, **7**(9), pp. 8578–8596.

27. Wittmar, A., Wittmar, M., Ulrich, A., Caparrotti, H., and Veith, M. (2012). Hybrid sol-gel coatings doped with transition metal ions for the protection of AA 2024-T3, *J. Sol-Gel Sci. Technol.*, **61**(3), pp. 600–612.

28. Poelman, M., Lahem, D., Atmani, F., Recloux, I., and Olivier, M. (2011). Protecting mild steel against corrosion in neutral aqueous solution using combined inhibitors: a classical and local electrochemical approach, *EUROCORR 2011: European Corrosion Congress*, **4**, p. 2822.

29. Lashgari, M. (2011). Theoretical challenges in understanding the inhibition mechanism of aluminum corrosion in basic media in the presence of some p-phenol derivatives, *Electrochim. Acta*, **56**(9), pp. 3322–3327.

30. Lahem, D., Poelman, M., and Olivier, M. (2010). Synergistic improvement of the inhibitory activity of dicarboxylates in preventing mild steel corrosion in neutral aqueous solution, *EUROCORR 2010: European Corrosion Congress*, **3**, p. 1897.

31. Madkour, L. H., and Zinhome, U. A. (2010). Inhibition effect of Schiff base compounds on the corrosion of iron in nitric acid and sodium hydroxide solutions, *J. Corros. Sci. Eng.*, **13**.

32. Afshari, V., and Dehghanian, C. (2010). The influence of nanocrystalline state of iron on the corrosion inhibitor behavior in aqueous solution, *J. Appl. Electrochem.*, **40**(11), pp. 1949–1956.

33. Qiao, G.-M., Ren, Z.-J., Zhang, J., et al.　(2010). Molecular dynamics simulation of corrosive medium diffusion in corrosion inhibitor membrane, *Wuli Huaxue Xuebao/Acta Physico – Chimica, Sinica*, **26**(11), pp. 3041–3046.

34. Rao, B. V. A., Rao, M. V., Rao, S. S., and Sreedhar, B. (2010). Tungstate as a synergist to phosphonate-based formulation for corrosion control of carbon steel in nearly neutral aqueous environment, *J. Chem. Sci.*, **122**(4), pp. 639–649.

35. Liu, L.-F., Liu, J.-X., Zhang, J., et al. (2010). Molecular dynamics simulation of the corrosive medium diffusion behavior inhibited by the corrosion inhibitor membranes, *Gaodeng Xuexiao Huaxue Xuebao/Chemical Journal of Chinese Universities*, **31**(3), pp. 537–541.

36. Finšgar, M., Fassbender, S., Hirth, S., and Milošev, I. (2009). Electrochemical and XPS study of polyethyleneimines of different molecular sizes as corrosion inhibitors for AISI 430 stainless steel in near-neutral chloride media, *Mater. Chem. Phys.*, **116**(1), pp. 198–206.

37. El-Rabiee, M. M., Helal, N. H., El-Hafez, Gh. M. A., and Badawy, W. A. (2008). Corrosion control of vanadium in aqueous solutions by amino acids, *J. Alloys Compd.*, **459**(1–2), pp. 466–471.

38. Al-Taher, F., Telegdi, J., and Kálmán, E. (2007). Effect of divalent cations in Langmuir-Blodgett films on the protection of copper against corrosion, *Mater. Sci. Forum*, **537–538**, pp. 9–16.

39. Wahbi, Z., Guenbour, A., Abou El Makarim, H., Ben Bachir, A., and El Hajjaji, S. (2007). Study of the inhibition of the corrosion of iron steel in neutral solution by electrochemical techniques and theoritical calculations, *Prog. Org. Coat.*, **60**(3), pp. 224–227.

40. Mesbah, A., Juers, C., Lacouture, F., et al. (2007). Inhibitors for magnesium corrosion: metal organic frameworks, *Solid State Sci.*, **9**(3–4), pp. 322–328.

41. Ravichandran, R. (2006). Protective effect of new class of benzotriazole derivatives towards corrosion of brass in neutral medium, *J. Corros. Sci. Eng.*, **10**.

42. Natarajan, S., and Kumaresh Babu, S.P. (2006). Corrosion and its inhibition in SA213-T22 TIG weldments used in power plants under neutral and alkaline environments, *Mater. Sci. Eng. A*, **432**(1–2), pp. 47–51.

43. Berchmans, L. J., Sivan, V., and Iyer, S. V. K. (2006). Studies on triazole derivatives as inhibitors for the corrosion of muntz metal in acidic and neutral solutions, *Mater. Chem. Phys.*, **98**(2–3), pp. 395–400.

44. Savchenko, O. N., Sizaya, O. I., and Gumenyuk, O. L. (2005). Use of modified mustard oil in steel corrosion protection, *Prot. Met.*, **41**(6), pp. 573–580.

45. Girčiene, O., Samulevičiene, M., Sudavičius, A., and Ramanauskas, R. (2005). Efficiency of corrosion inhibitors for steel in alkaline solutions and concrete structures, *Bull. Electrochem.*, **21**(7), pp. 325–330.

46. El-Etre, A. Y., Abdallah, M., and El-Tantawy, Z. E. (2005). Corrosion inhibition of some metals using lawsonia extract, *Corros. Sci.*, **47**(2), pp. 385–395.

47. Gunasekaran, G., Dubey, B. I., and Rangarajan, J. (2005). Role of citrate ions in the phosphonate-based inhibitor system for mild steel in aqueous chloride media, *Defence Sci. J.*, **55**(1), pp. 51–62.

48. Ogurtsov, N. A., Pud, A. A., Kamarchik, P., and Shapoval, G. S. (2004). Corrosion inhibition of aluminum alloy in chloride mediums by undoped and doped forms of polyaniline, *Synth. Met.*, **143**(1), pp. 43–47.

49. Cruz, J., Martínez, R., Genesca, J., and García-Ochoa, E. (2004). Experimental and theoretical study of 1-(2-ethylamino)-2-methylimidazoline as an inhibitor of carbon steel corrosion in acid media, *J. Electroanal. Chem.*, **566**(1), pp. 111–121.

50. Subramania, A., Sundaram, N. T. K., Priya, R. S., Muralidharan, V. S., and Vasudevan, T. (2004). Polymeric corrosion inhibitors: an overview, *Bull. Electrochem.*, **20**(2), pp. 49–58.

Chapter 11

Corrosion Inhibitors for Metals and Alloys in an Alkaline Medium

Metals undergo corrosion in an acid medium, a basic medium, and a neutral medium. In all the mediums, the anodic reaction is dissolution of the metal. In an acidic medium, the cathodic reaction is evolution of hydrogen. In a neutral medium and in an alkaline medium, the cathodic reaction is the formation of hydroxide ions. The corrosion inhibitor for an alkaline medium is discussed in this chapter.

Many inhibitors have been used to control corrosion of metals in a neutral medium [1–20]. Metals such as mild steel [1], copper [1, 8], aluminum and its alloys [2, 5–8], iron [4], and zinc [20] have been used in alkaline mediums. Corrosion inhibitors such as tender areca nut husk extract [1], cactus mucilage and brown seaweed extract [3], fruits of *Terminalia chebula* [5], Schiff bases [12], and o-benzoic acid derivatives have been used as corrosion inhibitors.

Many methods have been used to evaluate the corrosion inhibition of metals. Methods such as the weight loss method [1, 2] and electrochemical studies have been used [2, 4, 10]. During corrosion inhibition, protective films are formed on the metal surface. They usually consist of metal inhibitor samples. The surface

Titanic Corrosion
Susai Rajendran and Gurmeet Singh
Copyright © 2020 Jenny Stanford Publishing Pte. Ltd.
ISBN 978-981-4800-95-2 (Paperback), 978-1-003-00449-3 (eBook)
www.jennystanford.com

is analyzed by various surface analysis techniques such as scanning electron microscopy (SEM) [1, 5, 9, 11], atomic force microscopy (AFM) [2, 5], energy-dispersive X-ray spectroscopy (EDX) [5], X-ray diffraction (XRD) [8], and Fourier transform infrared spectroscopy (FTIR) [17].

Corrosion inhibition studies have been evaluated at various temperatures [12, 13, 18–20]. In general, corrosion inhibition decreases because at elevated temperature, desorption of adsorbed molecules takes place. Adsorption isotherm studies have also been carried out. The Temkin isotherm [4, 12] and Langmuir isotherm [5, 16] have been proposed.

Various corrosion inhibitors for metals in an alkaline medium are summarized in Table 11.1.

11.1 Case Study: Inhibition of Corrosion of Aluminum (6061) Immersed in Aqueous Solution at pH 11

The protective film was analyzed by Fourier transform infrared spectroscopy (FTIR) spectra and fluorescence spectra. The surface morphology was studied by AFM. HOMO-LUMO treatment was also employed.

A formulation consisting of 250 ppm p-nitrophenol and 50 ppm of Zn^{2+} offers a corrosion inhibition efficiency of 90% in controlling the corrosion of aluminum (6061) immersed in aqueous solution at pH 11. The polarization study (Fig. 11.1 and Table 11.2) reveals that the inhibitor system functions as an anodic inhibitor, controlling the anodic reaction predominantly. There is an increase in the linear polarization resistance (LPR) value and a decrease in corrosion current. This indicates that a protective film is formed on the metal surface, which prevents the transfer of electrons from metal to medium. Further AC impedance data (Fig. 11.2, Fig. 11.3, and Table 11.3) confirm the formation of a protective film. In the presence of an inhibitor system, charge transfer resistance increases and double-layer capacitance decreases.

Table 11.1 Corrosion inhibitors for metals in an alkaline medium

S. no.	Metal/Medium	Inhibitor/ Additives	Methods	Findings	Refs.
1	Mild steel and copper metals in both 0.5 M HCl and 0.5 M NaOH media	Tender areca nut husk (green color) extract	Weight loss measurements, Tafel plot and AC impedance studies, SEM	93.75% for copper in an acidic medium, 90% for copper in an alkaline medium, 93.478% for mild steel in an acidic medium, and 91.66% for mild steel in an alkaline medium.	1
2	Aluminum in both 0.5 M HCl and 0.1 M NaOH environments	Areca nut husk extracts	Weight loss measurements, Tafel plot, XRD, SEM, AFM	The inhibited aluminum surface is superior compared to the uninhibited aluminum surface.	2
3	Steel corrosion in chloride-contaminated alkaline media	Cactus mucilage and brown seaweed extract	Electrochemical cells, linear polarization resistance, and electrochemical impedance spectroscopy (EIS) measurements	Both additions reduce the corrosion rate of rebars and pitting in an alkaline medium with chloride ions (16 g/L NaCl).	3

(Continued)

Table 11.1 (*Continued*)

S. no.	Metal/Medium	Inhibitor/ Additives	Methods	Findings	Refs.
4	Iron in acidic and alkaline media	Polydentate Schiff base compounds	Weight loss, thermometric, potentiodynamic polarization (PDP) measurements and quantum chemical study, highest occupied molecular orbital (HOMO), lowest unoccupied molecular orbital (LUMO)	The high inhibition efficiency of the inhibitor was attributed to the blocking of active sites by adsorption of inhibitor molecules on the iron surface. It obeyed the Temkin isotherm.	4
5	6063 aluminum alloy in sodium hydroxide solution	Fruits of *T. chebula*	Weight loss method and PDP method, SEM, EDX	Inhibition efficiency increased with concentration of the inhibitor and decreased with temperature. The adsorption obeyed the Langmuir adsorption isotherm.	5
6	7A04 aluminum alloy	2nd International Conference on Mechanical Design and Power Engineering, ICMDPE 2013	Effect of aging treatments on the tensile strength and stress corrosion cracking (SCC) resistance of 7A04 aluminum alloy study	Performance evaluation of the carbon dioxide corrosion inhibitor IMC-1.	6

S. no.	Metal/Medium	Inhibitor/ Additives	Methods	Findings	Refs.
7	6061 Al-15 vol. pct. SiC(p) composite/0.5-M sodium hydroxide (NaOH) solution	4-amino-5-phenyl-4H-1,2,4-triazole-3-thiol (APTT)	PDP and EIS techniques	The inhibition efficiency of APTT increases with the increase in the concentration of the inhibitor and decreases with the increase in temperature.	7
8	Copper (99.9%), aluminum alloy (AA 6061), and carbon steel (SAE 1045)/ sodium hydroxide (NaOH)	Silicate	XRD analysis, X-ray fluorescence (XRF) analysis	Determine the appropriate SiO_2:Na_2O ratio and understand how this SiO_2:Na_2O ratio can affect the corrosion rate of each metal alloys immersed in an acidic medium.	8
9	6061-Al alloy/aqueous sodium hydroxide solution	3-Ethyl-4-amino-5-mercapto-1,2,4-triazole (EAMT)	SEM	The presence of EAMT in sodium hydroxide solution decreases the corrosion rates and the corrosion current densities (I_{corr}) and increases the charge transfer resistance (R_p).	9
10	6061/Al-15 vol% SiCp composite in different concentrations of sodium hydroxide solution	EAMT	EIS techniques	The inhibitor efficiency depends on the concentration of the inhibitor; the concentration of the corrosive media, and temperature.	10

(Continued)

Table 11.1 (*Continued*)

S. no.	Metal/Medium	Inhibitor/Additives	Methods	Findings	Refs.
11	6061 Al alloy in 0.5 M sodium hydroxide	3-Methyl-4-amino-5-mercapto-1,2,4-triazole (MAMT)	PDP and EIS techniques, SEM	The inhibition efficiency increased with the increase in the concentration of the inhibitor but decreased with the increase in temperature.	11
12	Iron electrode/2.0 mol/L nitric acid and 2.0 mol/L sodium hydroxide solutions	Schiff bases	Chemical and electrochemical techniques, adsorption isotherm	The adsorbed Schiff base molecules interact with iron ions formed on the metal surface, leading to neutral and cationic iron–Schiff base complexes; Temkin isotherm.	12
13	Aluminum alloy (AA3003) in aqueous solutions of sodium hydroxide	*Euphorbia hirta* and *Dialum guineense* extracts	Weight loss technique, temperature study	The higher the temperature, the lower the inhibition efficiency, which is due to the fact that the rate of corrosion of aluminum is higher than the rate of adsorption.	13

S. no.	Metal/Medium	Inhibitor/ Additives	Methods	Findings	Refs.
14	Aluminum/NaOH solution	Thiourea, phenyl thiourea (PTU) and 4-carboxy phenyl thiourea (CPTU)	Weight loss technique	The extent of adsorption depends on the nature of the metal, the metal surface condition, the mode of adsorption, the chemical structure of the inhibitor, and the type of corrosive media.	14
15	Aluminum in aqueous solutions of sodium hydroxide	Sulfonic acid, sodium cumene sulfonate, and sodium alkyl sulfate	Weight loss technique, temperature study	Inhibition efficiency of the inhibitors tested increases with decreasing sodium hydroxide concentrations. The higher the temperature, the lower the inhibition efficiency, which is due to the fact that the rate of corrosion of aluminum is higher than the rate of adsorption.	15
16	Aluminum (Al)/ hydrochloric acid (HCl) and sodium hydroxide (NaOH) solutions	Cetyl trimethylammonium chloride (CTAC)	Weight loss tests and potentiostatic polarization measurements	In both acidic and basic media, the increase in temperature resulted in a decrease of the inhibition efficiency and a decrease in the degree of surface coverage; Langmuir isotherm in NaOH.	16

(Continued)

Table 11.1 (*Continued*)

S. no.	Metal/Medium	Inhibitor/ Additives	Methods	Findings	Refs.
17	Mild steel immersed in an aqueous solution containing 60 ppm of Cl, caffeine-Zn^{2+}	Sodium sulfite (Na_2SO_3), sodium dodecyl sulfate (SDS)	FTIR	Formation of micelles by surfactants changes the inhibition efficiency.	17
18	Aluminum in aqueous solutions of sodium hydroxide	Sulfonic acid, sodium cumene sulfonate, and sodium alkyl sulfate	Weight loss technique, temperature study	The higher the temperature, the lower the inhibition efficiency.	18
19	Low-carbon steel and certain aluminum alloys in acidic and in alkaline media	Wastes from processing of the White Sea algae *Fucus*	Temperature study	Extracts of algae used as inhibitor for carbon steel and aluminum alloys.	19
20	Zinc in 2 M NaOH	o-Benzoic acid derivatives	Thermometric and weight loss methods	The results indicate that the additives lie flat on the surface of the corroding metal.	20

Table 11.2 Corrosion parameters of aluminum (6061) immersed in aqueous solution (pH 11, NaOH) in the presence and absence of an inhibitor system obtained by the polarization study

System	E_{corr} mV vs. SCE	b_c (mV/ decade)	b_a (mV/ decade)	LPR (ohm·cm^2)	I_{corr} (A/cm^2)
Blank	−1573	180	641	215	2.848×10^{-4}
p- Nitrophenol (250 ppm) + Zn^{2+} (50 ppm)	−1479	162	214	1546	2.952×10^{-5}

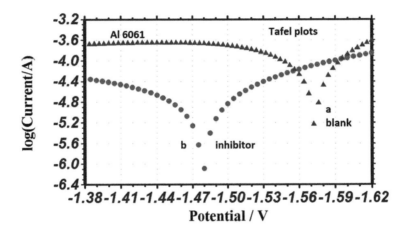

Figure 11.1 Polarization curves of Al 6061 immersed in various test solutions at pH 11.

Table 11.3 Corrosion parameters of Al 6061 in various test solutions obtained from AC impedance spectra

System	R_t (ohm·cm^2)	C_{dl} (F/cm^2)	Impedance log (Z/ohm)
Blank	50.01	10.198×10^{-8}	1.912
p-Nitrophenol (250 ppm) + Zn^{2+} (50 ppm)	322	1.584×10^{-8}	2.736

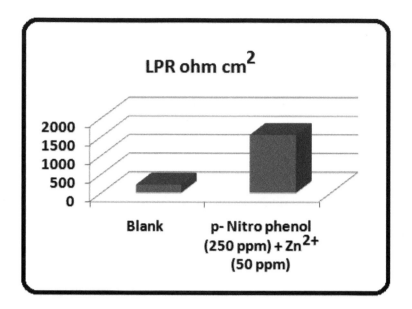

Figure 11.2 Comparison of LPR values.

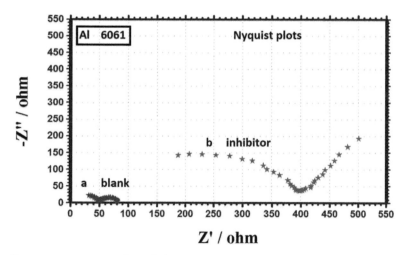

Figure 11.3 Nyquist plots of Al 6061 in various test solutions at pH 11 obtained from AC impedance spectra.

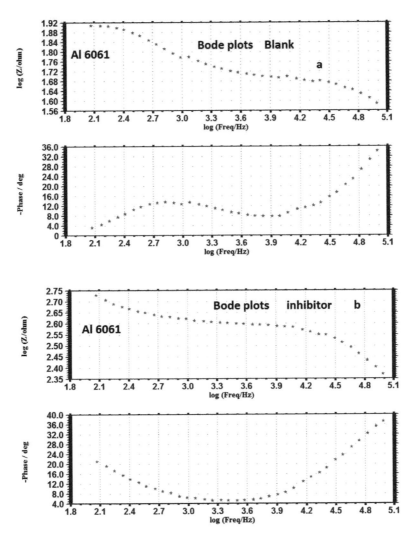

Figure 11.4 Bode plots of Al 6061 in (a) a blank solution and (b) an inhibitor solution.

11.1.1   Analysis of Fluorescence Spectra

The fluorescence spectrum (λ_{ex} = 380 nm) of the film formed on the metal surface after immersion in the solution containing p-nitrophenol and Zn^{2+} reveals that a peak appears at 400.5 nm.

This peak is due to the p-nitrophenol–Al^{3+} complex formed on the metal surface. This peak matches with the peak (400.0 nm) of the p-nitrophenol–Al^{3+} complex formed in the solution. Hence it is concluded that the protective film consists of a p-nitrophenol–Al^{3+} complex formed on the metal surface. Since only one peak is obtained, the complex is highly symmetrical and homogeneous. In the fluorescence spectrum of the protective film, there is a decrease in intensity. This is due to the fact that the film is in the solid state, where the electron transition is restricted and difficult.

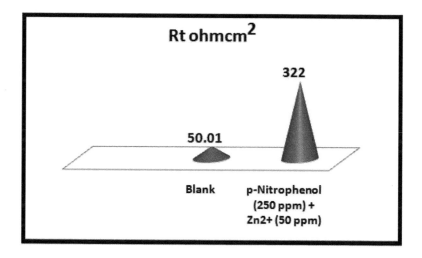

Figure 11.5 Comparison of R_t (ohm·cm^2) values.

11.1.2 FTIR Spectra

FTIR spectra have been used to analysis the protective film formed on the metal surface. The FTIR spectrum of p-nitrophenol is shown in Fig. 11.6. The broad band at 3325 cm^{-1} observed in the spectrum of p-nitrophenol indicates the –OH stretching vibration. A peak at 1590 cm^{-1} is attributed for the –N=O stretching vibration. The C=O stretching frequency of the phenolic compound occurs at 1105 cm^{-1}.

The FTIR spectrum of the protective film formed on the surface of aluminum (6061) immersed in p-nitrophenol (250 ppm) and Zn^{2+} (50 ppm) at pH 11 is shown in Fig. 11.7. The –OH stretching

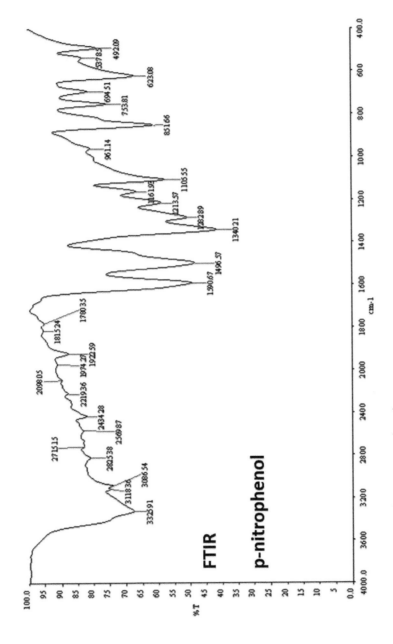

Figure 11.6 FTIR spectrum of pure p-nitrophenol.

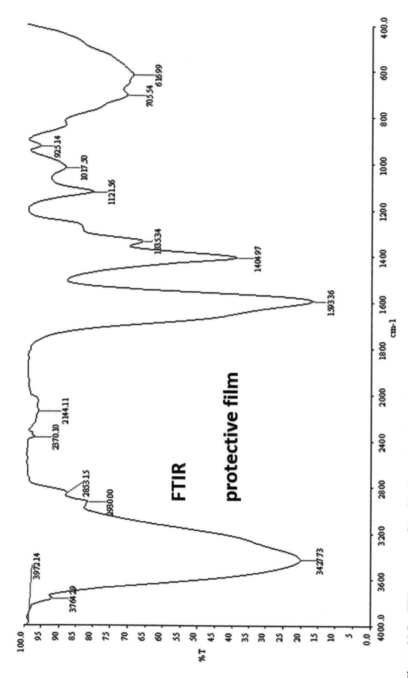

Figure 11.7 FTIR spectrum of the film formed on the surface of aluminum (6061) immersed in p-nitrophenol (250 ppm) and Zn^{2+} (50 ppm) at pH 11.

frequency has increased from 3325 cm^{-1} to 3427 cm^{-1}. This shows that Al^{3+} has coordinated through oxygen atom of the –OH group in p-nitrophenol. The –N=O stretching frequency has shifted from 1590 cm^{-1} to 1593 cm^{-1}. The C=O stretching frequency has increased from 1105 cm^{-1} to 1121 cm^{-1}. So p-nitrophenol has coordinated with Al^{3+} through the phenolic oxygen, resulting in the formation of an Al^{3+}–p-nitrophenol complex. The band at 616 cm^{-1} may be due to the metal–oxygen stretching vibration, and the band at 1404 cm^{-1} may be due to in-plane vibrations of the –OH group in Zn(OH)$_2$. This indicates the presence of Zn(OH)$_2$ on the metal surface. There is also possibility of the formation of an Al(OH)$_3$ precipitate on the metal surface.

When aluminum (6061) is immersed in this solution the Zn^{2+}–p-nitrophenol complex diffuses from the bulk of the solution toward the metal surface. The Zn^{2+}–p-nitrophenol complex is converted into an Al^{3+}–p-nitrophenol complex on the metal surface. Zn^{2+} is released and Zn(OH)$_2$ is formed. Thus a protective film consisting of an Al^{3+}–p-nitrophenol complex is formed on the anodic metal surface.

11.1.3 AFM Characterization

AFM (Figs. 11.8–11.10 and Table 11.4) reveals the formation of nanofilms on the metal surface in the presence of inhibitors. The root-mean-square (RMS) roughness value, average roughness value, and peak-to-valley heights have been calculated for a polished metal surface, a polished metal surface immersed in a corrosive medium (pH 11), and a polished metal surface immersed in an inhibitor system. It is observed that in the presence of an inhibitor system the RMS roughness value is higher than that of the polished metal surface but lower than that of the polished metal surface immersed in a corrosive medium (pH 11). Similar is the case with other parameters also, namely average roughness value and peak-to-valley heights. Thus AFM leads to the conclusion that self-assembling nanofilms of p-nitrophenol are formed on the metal surface, which control the corrosion process.

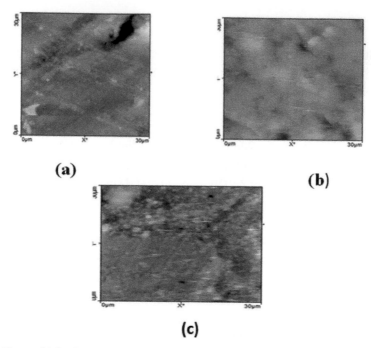

Figure 11.8 2D AFM images of the surface of (a) polished aluminum, (b) aluminum immersed in an aqueous solution at pH 11 (blank), and (c) aluminum immersed in a solution containing p-nitrophenol (250 ppm) and Zn^{2+}(50 ppm) at pH 11.

Table 11.4 AFM data for aluminum (6061) surface immersed in an uninhibited environment

Sample	RMS roughness R_q (nm)	Average roughness R_a (nm)	Maximum peak-to-valley height (nm)
Polished aluminum	41.46	32.56	132.41
Aluminum immersed in aqueous solution at pH 11 (blank)	267.62	239.13	948.62
Aluminum immersed in aqueous solution at pH 11 + p-nitrophenol (250 ppm) + Zn^{2+} (50 ppm)	80.769	61.543	289.98

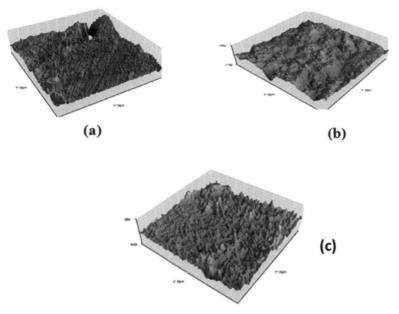

(a)

(b)

(c)

Figure 11.9 3D AFM images of the surface of (a) polished aluminum, (b) aluminum immersed in an aqueous solution at pH 11 (blank), and (c) aluminum immersed in a solution containing p-nitrophenol (250 ppm) and Zn^{2+}(50 ppm) at pH 11.

11.1.4 Density Functional Theory: Computational Methods

Accurate information about geometrical configuration, electron distribution, and quantum chemical calculations were provided by density functional theory (DFT), which is an economic and efficient quantum chemistry computing method. It is widely applied in the analysis of corrosion inhibition performance and the interaction of corrosion inhibitors and interfaces. The B3LYP functional was applied within the context of Gaussian 09 using the 6-31+G(d, p) basis set. From a quantum chemical study, the mechanism of corrosion, inhibitor adsorption on metal surfaces, and the merits of corrosion inhibitors can easily be studied. There is a correlation between experimental results and the quantum chemical parameters computed using DFT. The following quantum chemical indices were taken into consideration: energy of the HOMO (E_{HOMO}), energy of the LUMO (E_{LUMO}), energy band gap (ΔE), electron affinity (A), ionization potential (I), hardness (η), and softness ($1/\eta$).

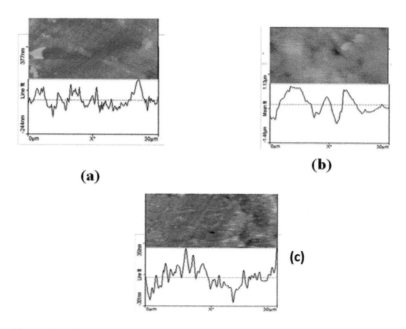

(a)

(b)

(c)

Figure 11.10 AFM cross-sectional images of the surface of (a) polished aluminum, (b) aluminum immersed in an aqueous solution at pH 11 (blank), and (c) aluminum immersed in a solution containing p-nitrophenol (250 ppm) and Zn^{2+} (50 ppm) at pH 11.

The calculations were done by using the Gaussian 09 program. Schematic structures were drawn using the Chem Office package in the Ultra Chem 2010 version, while optimized structures were drawn using the Gauss View 5 program.

The frontier molecule orbital density distributions of p-nitrophenol are presented in Fig. 11.11. The calculated quantum chemical parameters are given in Table 11.5.

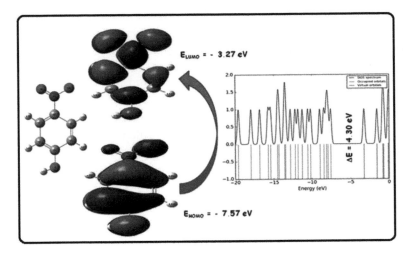

Figure 11.11 Schematic representation of HOMO and LUMO of p-nitrophenol.

Table 11.5 Calculated quantum chemical and corrosion parameters of the inhibitor p-nitrophenol

Parameters	Value
E_{HOMO}	−7.57eV
E_{LUMO}	−3.27eV
ΔE ($E_{\mathrm{LUMO}} - E_{\mathrm{HOMO}}$)	4.30 eV
I ($-E_{\mathrm{HOMO}}$)	7.57 eV
A ($-E_{\mathrm{LUMO}}$)	3.27 eV
η ($I - A/2$)	2.15
Softness ($1/\eta$)	0.4651
Inhibition efficiency (%): weight loss method	90
LPR (ohm·cm^2)	1546
I_{corr} (A/cm^{-2})	2.952×10^{-5}
R_{t} (ohm·cm^2)	322
C_{dl} (F/cm^2)	1.584×10^{-8}
Impedance value log(Z/ohm)	2.736
R_{q} (nm)	80.769
R_{a} (nm)	61.543
Maximum peak-to-valley height (nm)	289.98

The inhibition effect is due to the adsorption of the inhibitor molecule on the metal surface. Adsorption may be physical adsorption (physisorption) or chemical adsorption (chemisorption), depending on the adsorption strength. In chemisorption, one of the reacting species acts as an electron donor and the other one acts as an electron acceptor. The energy of the HOMO (E_{HOMO}) measures the tendency toward the donation of an electron by a molecule. High values of E_{HOMO} have a tendency of the molecule to donate electrons to appropriate acceptor molecules.

An inhibitor is not only an electron donor but also an electron acceptor. It can accept an electron from the d-orbital of the metal, leading to the formation of a feedback bond. The tendency for the formation of a feedback bond depends on the value of E_{LUMO}. The lower the E_{LUMO}, the easier the acceptance of electrons from the d-orbital of the metal.

Organic compounds not only offer electrons to the unoccupied metal orbital but also accept free electrons from the metal. A molecule with a low energy gap is more polarizable and is generally associated with high chemical activity and low kinetic stability and is termed "soft molecule." A soft molecule is more reactive than a hard molecule because a hard molecule has a large energy gap.

Using the DFT/B3LYP method, the inhibition efficiency of a p-nitro molecule was investigated.

11.2 Conclusion

The p-nitrophenol–Zn^{2+} system offers good protection to Al 6061 immersed in an aqueous solution at pH 11. The protective film consists of an aluminum–p-nitrophenol complex. The protective film is in the nanometer scale.

References

1. Raghavendra, N., and Ishwara Bhat, J. (2019). Application of green products for industrially important materials protection: an amusing anticorrosive behavior of tender arecanut husk (green color) extract at metal-test solution interface, *Measurement*, **135**, pp. 625–639.

2. Raghavendra, N., and Ishwara Bhat, J. (2018). Benevolent behavior of arecanut husk extracts as potential corrosion inhibitor for aluminum

in both 0.5 M HCl and 0.1 M NaOH environments, *J. Bio- Tribo-Corros.*, **4**(3), p. 44.

3. Hernández, E. F., Cano-Barrita, P. F. D. J., León-Martínez, F. M., and Torres-Acosta, A. A. (2017). Performance of cactus mucilage and brown seaweed extract as a steel corrosion inhibitor in chloride contaminated alkaline media, *Anti-Corros. Methods Mater.*, **64**(5), pp. 529–539.

4. Madkour, L. H., and Elroby, S. K. (2015). Inhibitive properties, thermodynamic, kinetics and quantum chemical calculations of polydentate Schiff base compounds as corrosion inhibitors for iron in acidic and alkaline media, *Int. J. Ind. Chem.*, **6**(3), pp. 165–184.

5. Prabhu, D., and Rao, P. (2015). Inhibiting action of fruits of Terminalia chebula on 6063 aluminum alloy in sodium hydroxide solution, *J. Mater. Environ. Sci.*, **6**(2), pp. 412–424.

6. ICMDPE 2013 (2014). 2nd International Conference on Mechanical Design and Power Engineering, *Appl. Mech. Mater.*, pp. 490–491.

7. Reena Kumari, P. D., Nayak, J., and Shetty, A. N. (2014). Corrosion inhibition of aluminum type 6061 Al-15 vol. pct. SiC(p) composite in 0.5-M sodium hydroxide solution by 4-amino-5-phenyl-4H-1,2,4-triazole-3-thiol, *Anti-Corros. Methods Mater.*, **61**(4), pp. 241–249.

8. Mohamad, N., Othman, N. K., and Jalar, A. (2013). Investigation of $SiO_2:Na_2O$ ratio as a corrosion inhibitor for metal alloys, *AIP Conf. Proc.*, **1571**, pp. 136–140.

9. Reena Kumari, P. D., Nayak, J., and Nityananda Shetty, A. (2011). 3-ethyl-4-amino-5-mercapto-1,2,4-triazole as corrosion inhibitor for 6061-alloy in sodium hydroxide solution, *Port. Electrochim. Acta*, **29**(6), pp. 445–462.

10. Kumari, P. D. R., Nayak, J., and Shetty, A. N.(2011). 3-Ethyl-4-amino-5-mercapto-1,2,4-triazole as a corrosion inhibitor for 6061/Al-15 vol% SiCp composite in a sodium hydroxide solution, *Synth. React. Inorg. Met.-Org. Chem.*, **41**(7), pp. 774–784.

11. Kumari, P. D. R., Nayak, J., and Shetty, A. N.(2011). 3-Methyl-4-amino-5-mercapto-1,2,4-triazole as corrosion inhibitor for 6061 Al alloy in 0.5 M sodium hydroxide solution, *J. Coat. Technol. Res.*, **8**(6), pp. 685–695.

12. Madkour, L. H., and Zinhome, U. A. (2010). Inhibition effect of Schiff base compounds on the corrosion of iron in nitric acid and sodium hydroxide solutions, *J. Corros. Sci. Eng.*, **13**.

13. Nnanna, L. A., Onwuagba, B. N., Mejeha, M. I., and Okeoma, K. B. (2010). The effects of some plant extracts on corrosion and adsorption mechanism process of aluminium in alkaline solution, *J. Corros. Sci. Eng.*, **13**.

14. Edrah, S., and Hasan, S. K. (2010). Studies on thiourea derivatives as corrosion inhibitor for aluminum in sodium hydroxide solution, *J. Appl. Sci. Res.*, **6**(8), pp. 1045–1049.

15. Maayta, A. K. (2006). Organic corrosion inhibitors for aluminium in sodium hydroxide, *Asian J. Chem.*, **18**(1), pp. 566–572.

16. Al-Rawashdeh, N. A. F., and Maayta, A. K. (2005). Cationic surfactant as corrosion inhibitor for aluminum in acidic and basic solutions, *Anti-Corros. Methods Mater.*, **52**(3), pp. 160–166.

17. Rajendran, S., Vaibhavi, S., Anthony, N., and Trivedi, D. C. (2003). Transport of inhibitors and corrosion inhibition efficiency, *Corrosion*, **59**(6), pp. 529–534.

18. Maayta, A. K. (2002). Organic corrosion inhibitors for aluminum in sodium hydroxide, *J. Corros. Sci. Eng.*, **3**(25).

19. Popelyukh, G. M., and Talavira, L. I. (1986). Marine vegetable products as inhibitors of metal corrosion, *J. Appl. Chem. USSR*, **59**(11 pt 2), pp. 2363–2365.

20. Fouda, A. S. (1982). Effect of o-benzoic acid derivatives on the corrosion behaviour of zinc in 2 M sodium hydroxide solution with and without barium, *Indian J. Technol.*, **20**(10), pp. 412–415.

Chapter 12

Corrosion Inhibition of Metals and Alloys in the Presence of Extracts of Plant Materials: An Overview

Extracts of plant materials are low cost and environmentally friendly. They contain many active ingredients. The molecules of these ingredients may contain hetero- or polar atoms such as sulfur (S), nitrogen (N), oxygen (O), and phosphorous (P). The lone pair of electrons present in these atoms is pumped onto the metal surface. Thus the loss of electrons from the metal surface can be avoided and corrosion inhibition takes place. Because of the adsorption of inhibitor molecules on the metal surface, a protective film is formed on the metal surface and corrosion is controlled.

12.1 Introduction

Corrosion is a spontaneous and thermodynamically favorable process. Because of corrosion, every year 5% gross national product (GNP) of any country is lost. However, 50% of corrosion loss can be avoided using available knowledge. There are many methods used to control corrosion. One such method is the use of inhibitors. Once chromates were used as corrosion inhibitors. However,

Titanic Corrosion
Susai Rajendran and Gurmeet Singh
Copyright © 2020 Jenny Stanford Publishing Pte. Ltd.
ISBN 978-981-4800-95-2 (Paperback), 978-1-003-00449-3 (eBook)
www.jennystanford.com

environmental scientists do not support this, because chromates are toxic in nature. So researchers are looking for environmentally friendly corrosion inhibitors such as extracts of plant materials. Several plant materials have been used for this purpose [1–84].

Extracts of plant materials contain many active ingredients. They contain polar atoms such as S, N, O, P, etc. Because of this nature, the lone pair of electrons present in these atoms is pumped onto the metal surface. Thus the loss of electrons from the metal surface can be avoided. Hence corrosion inhibition takes place. Because of the adsorption of inhibitor molecules on the metal surface, a protective film is formed. Thus corrosion is controlled. Various extracts of plant materials have been used to prevent a variety of metals immersed in various mediums at different temperatures in the presence and absence of many additives. Many methods have been employed (such as weight loss method, electrochemical studies, etc.) to evaluate corrosion inhibition efficiencies of inhibitors. The protective film has been analyzed by various surface analyses techniques.

Extracts of plant materials have been used to control corrosion of various metals such as steel [15, 18], mild steel [1–9], aluminum and its alloys [11, 12, 27], zinc [25, 30], and copper [42, 52]. Extracts of plant materials have also been used to prevent corrosion of metals and alloys immersed in various mediums such as acidic mediums [1, 3, 5–7], basic mediums [27, 29, 31], and neutral mediums [60, 64, 70]. Experiments have been conducted at various temperatures [8, 9, 14]. Extracts of various parts of plants are used as corrosion inhibitors. Fruits [37], leaves [1, 5, 8, 15], roots [49], bark [42], flowers [61], seeds [11], stems [56], gum [42], husk [33], and peel [45, 63] have been used as corrosion inhibitors.

A mixture of inhibitors shows better inhibition efficiency than individual members. This is called the synergistic effect. For this purpose many additives have been used to improve the inhibition efficiency of plant extracts. For this purpose, halide [44], sulfide [52], and phosphonic acid [59] have been used.

To evaluate the corrosion inhibition efficiency of various plant extracts, several methods such as weight loss method [1–4, 6] and electrochemical studies (polarization, AC impedance spectra) [8, 11, 14, 15] have been employed.

The protective films formed on metal surfaces have been analyzed by various surface analysis techniques, like Fourier transform infrared spectroscopy (FTIR) [18, 34, 38], UV-visible spectroscopy [46, 50, 70, 71], scanning electron microscopy (SEM) [41], energy-dispersive X-ray analysis (EDAX) [27], and atomic force microscopy (AFM) [36]. The protective film is formed by adsorption of active ingredients of various plant extracts on the metal surface. The adsorption isotherms are usually Freundlich adsorption isotherm [49, 55], Temkin adsorption isotherm [25, 46, 48], and Langmuir adsorption isotherm [1, 2, 5, 6]. From the adsorption isotherms various thermodynamic parameters such as changes in free energy, entropy, enthalpy, and activation energy [2, 11, 18] have been calculated.

Plant materials are extracted by making use of alcohol [1, 14], acid [43], and water [2, 3].

12.2 Significance and Limitations of Plant Extracts

Plant extracts are environmentally friendly, nontoxic, cheap, and biodegradable. However, they are easily degradable by microorganisms. The decomposition of plant extracts can be prevented by addition of biocides such as sodium dodecyl sulfate (SDS) and *N*-cetyl-*N,N,N*-trimethyl ammonium bromide. Rajendran et al. have extensively used extracts of plant materials as corrosion inhibitors [60–84]. A list of plant materials that have been used as corrosion inhibitors is given in Table 12.1.

Table 12.1 Plant materials used as corrosion inhibitors

S. no.	Metal/Medium	Inhibitor/Additive	Method	Findings	Refs.
1	X60 carbon steel and 15% HCl solution	Ethanolic extracts of date palm leaves and seeds	Weight loss, electrochemical techniques, and photochemical screening	Inhibition efficiancy increased with increase in concentration of the extracts and temperature. Mixed-type inhibitors. Langmuir adsorption isotherm model.	1
2	Carbon steel and saline formation water	*Centaurea cyanus* aqueous extract	Weight loss measurements, electrochemical impedance spectroscopy (EIS) and potentiodynamic polarization techniques	Mixed-type inhibitor, Langmuir adsorption isotherm, thermodynamic activation parameters.	2
3	Mild steel and HCl aqueous solution	Gorse aqueous extract	Weight loss measurements, potentiodynamic polarization curves, EIS, and SEM	Reducing both anodic and cathodic current densities.	3
4.	Carbon steel API 5LX and hypersaline environments	*Azadirachta indica* (neem) extract	Weight loss, electrochemical studies, FTIR, X-ray diffraction (XRD)	Inhibit the corrosion process.	4

S. no.	Metal/Medium	Inhibitor/Additive	Method	Findings	Refs.
5.	Mild steel and 1 M H_2SO_4 solution	*Acacia cyanophylla* leaf extract	Polarization techniques and EIS	Mixed inhibitor, Langmuir isotherm model.	5
6.	Mild steel and HCl solution	*Alpinia galanga*	Weight loss method, SEM	Langmuir adsorption isotherm.	6
7.	Mild steel and 15% and 20% HCl solutions	Coal-tar distillation products (CTDPs) and the aqueous extract of ginger	Weight loss of the steel	Decrease of corrosion rate and increase of inhibition efficiency.	7
8.	Carbon steel and carbon dioxide–saturated chloride carbonate solution	Olive leaf extract at 25°C and 65°C	Linear polarization resistance technique, EIS, SEM, and FTIR	The adsorption of olive leaf extract on the carbon steel surface.	8
9.	Mild steel and 15 wt% HCl	Durum wheat temperature (20°C–60°C) and immersion time (5–24 h)	Weight loss, FTIR, SEM, EDAX, and surface profilometry	Langmuir adsorption isotherm.	9
10	J55 steel and CO_2-saturated 3.5 wt% NaCl solution	Tangerine peel extract	Electrochemical techniques	Mixed-type green inhibitor; the Langmuir isothermal absorption model, and the El-Awady dynamic model.	10
11	Aluminum metal and 0.5 M HCl	Dry areca nut seed extract	Weight loss, electrochemical techniques, AFM, XRD	Langmuir isotherm, thermodynamic functions.	11

(Continued)

Table 12.1 *(Continued)*

S. no.	Metal/Medium	Inhibitor/Additive	Method	Findings	Refs.
12	Aluminum alloy AA3003 and HCl	*Kola nitida* extract	Experimental and computational approach	Langmuir adsorption.	12
13	Carbon steel and acidic solution	*Eulychnia acida* phil. (cactaceae) extracts	Mass loss	Blocking active sites on the metal.	13
14	Mild steel and 1.0 M HCl	Methanolic extract of *Mentha suaveolens* at 313–343 K	Weight loss measurement, potentiodynamic polarization, and EIS	Langmuir isotherm, mixed-type corrosion inhibitor.	14
15	Steel and 0.5 M HCl	Lemon leaf extracts with saffron, almonds, guava leaves, and *Origanum majorana* extracts	Electrochemical polarization and EIS methods	Langmuir adsorption isotherm.	15
16	Steel structures, their alloys, and saline corrosive medium (3.5% of sodium chloride)	Vegetal species *Copaifera* L. and copaiba oil sample (1.0% as oil phase), (10.0% as surfactant), and water (89.0%)	Electrochemical method	Langmuir model isotherm.	16
17	Mild steel and 1 M H_2SO_4	Extract of *Zygophyllum album*	Weight loss, electrochemical polarization, EIS, and SEM studies	Mixed mode of inhibition, protective layer on the surface of mild steel.	17

S. no.	Metal/Medium	Inhibitor/Additive	Method	Findings	Refs.
18	Steel and 8 mol·L^{-1} H$_3$PO$_4$	Extracts of *Lawsonia inermis*	Galvanostatic polarization studies, SEM, FTIR	Thermodynamic parameters show strong interaction between the inhibitor and the steel surface.	18
19	Mild steel and acidic media	*Piper nigrum* L. extract	Weight loss measurements	Excellent anticorrosion properties in acid conditions.	19
20	Carbon steel (UNS G10200) and 1.0 mol·L^{-1} H$_2$SO$_4$	*Aniba canelilla*	Mass spectrometry, polarization curve, EIS, and weight loss method	Good corrosion inhibition of high concentration of inhibitor.	20
21	Carbon steel and 1 mol·L^{-1} HCl	*Ilex paraguariensis* extracts	Electrochemical studies	Mixed-type inhibitor, Langmuir adsorption isotherm.	21
22	Mild steel and 1 M HCl solution	Hexane extract of *Nigella sativa* L.	Weight loss measurement, polarization curves, and EIS	Langmuir adsorption isotherm.	22
23	Mild steel and 1 M HCl solution	*Nigella sativa* L. extract	Weight loss measurements, electrochemical polarization, and EIS methods	Mixed inhibitor; Langmuir adsorption isotherm.	23

(Continued)

Table 12.1 (*Continued*)

S. no.	Metal/Medium	Inhibitor/Additive	Method	Findings	Refs.
24	Carbon steel and HCl, H_2SO_4 solutions	Green tea extracts	Weight loss, potentiodynamic polarization, EIS, and SEM	Good anticorrosion properties.	24
25	Zinc and 0.2 M HCl solution	*Azadirachta indica* (neem) extract	Mass loss, electrochemical polarization and impedance methods, SEM	Mixed-type inhibitor, Temkin adsorption isotherm.	25
26	Steel and 1 M HCl solution	*Mentha pulegium* extract	Potentiodynamic polarization and EIS	Mixed-type inhibitor, Temkin adsorption isotherm.	26
27	Al-Mg-Si and simulated 3.5 % sodium chloride solution	Green roasted *Elaeis guineensis* oil	Linear potentiodynamic polarization and gravimetric method SEM/EDAX	Langmuir adsorption isotherm.	27
28	Mild steel and 1 M HCl solution	Aqueous extract of gummara of *Phoenix dactylifera* (AEGPD) and aqueous extract of *Lactuca* (AEL)	Weight loss measurements, thermodynamic functions of dissolution, activation energy and adsorption processes	Langmuir adsorption isotherm.	28

S. no.	Metal/Medium	Inhibitor/Additive	Method	Findings	Refs.
29	Mild steel and natural seawater	Snow crab (*Chionoecetes opilio*) peptidic extracts	Electrochemical measurements, SEM	Good corrosion inhibitor, adsorption on the metallic surface by the inhibition of microbially induced corrosion (MIC).	29
30	Zinc and 0.5 M HCl	*Achillea fragrantissima*	Weight loss, hydrogen evolution, and polarization measurements	Langmuir adsorption isotherm.	30
31	Mild steel and seawater	Snow crab (*Chionocete opilio*)	Electrochemical measurements	Formation of a protective film on metallic surfaces.	31
32	Mild steel and 1 M HCl, 0.5 M H_2SO_4 solutions	Date palm (*Phoenix dactylifera*) seed extracts	Weight loss and electrochemical methods	Mixed-type inhibitor, Langmuir adsorption isotherm.	32
33	Q235A steel and 1 M HCl	*Diospyros kaki* L.f husk extracts	Weight loss and otentiodynamic polarization techniques	Mixed-type inhibitors.	33
34	Aluminum and its alloys	Plant extracts and natural products	Electrochemical methods, FTIR, reverse-phase high-performance liquid chromatography (RP-HPLC), SEM, energy-dispersive X-ray spectroscopy (EDS)	The protective films formed by the inhibitors.	34

(Continued)

Table 12.1 (*Continued*)

S. no.	Metal/Medium	Inhibitor/Additive	Method	Findings	Refs.
35	C-steel and molar HCl solution	*Chenopodium ambrorsioides* extract (CAE)	Weight loss measurements, potentiodynamic polarization curves, and EIS	Cathodic inhibitors.	35
36	Mild steel and HCl solution	Chamomile (*Matricaria recutita*) extract (CE)	Electrochemical (polarization, EIS) and surface analysis (optical microscopy/AFM/SEM) techniques	Mixed-type corrosion inhibitor.	36
37	Low-carbon steel	Extracts of the fruit of *Capsicum frutescens*	Gravimetric, impedance, and polarization techniques	Anticorrosion and antimicrobial activity.	37
38	Mild steel and 1 M HCl and 0.5 M H_2SO_4	Acid extracts of *Piper guineense* (PG) leaves	Gravimetric, potentiodynamic polarization and EIS techniques, as well as SEM and FTIR	Chemisorption of the extract constituents on the metal surface.	38
39	Steel and 1 M HCl	Argan kernel extract (AKE) and cosmetic argan oil (CAO)	Weight loss measurements, potentiodynamic polarization, and EIS measurements	Langmuir adsorption isotherm.	39

S. no.	Metal/Medium	Inhibitor/Additive	Method	Findings	Refs.
40	Mild steel and 1 M HCl	*Cassia auriculata* extract	Weight loss measurement and polarization study	The inhibition efficiency increases with increase in extract concentration.	40
41	Titanium alloy and 3.5% NaCl	Red algal extracts	EIS, SEM	Langmuir adsorption isotherm.	41
42	Copper and molar HCl	The alcoholic extract of leaves, stem bark, and gum from the *Acacia senegal*	Mass loss method and thermometric method	Langmuir adsorption isotherm, good protection to copper against corrosion.	42
43	Mild steel and 1 M HCl, 0.5 M H$_2$SO$_4$	Acid extracts of *Ervatamia coronaria* leaves	Conventional weight loss and electrochemical measurements	Mixed-type inhibitors	43
44	Mild steel and acid	Plant extracts and halide	Electrochemical and non-electrochemical techniques	Synergistic effects increased the inhibition efficiency in the presence of halide additives.	44
45	Carbon steel and 1 M HCl solution	Aqueous extracts of fruit peels	EIS, potentiodynamic polarization curves, weight loss measurements, and surface analysis	Good corrosion inhibitor, Langmuir adsorption isotherm.	45

(Continued)

Table 12.1 *(Continued)*

S. no.	Metal/Medium	Inhibitor/Additive	Method	Findings	Refs.
46	Pure copper, Cu-27 Zn brass and natural seawater	*Vitis vinifera* seed and skin extract	UV, IR, and XRD	Langmuir and Temkin adsorption isotherms.	46
47	Mild steel and 1 M HCl	*Eclipta prostrata* linn extract	Weight loss measurement and polarization study	Inhibition efficiency increases with increase in inhibitor concentration.	47
48	Mild steel and acid medium	*Cyamopsis tetragonaloba* seed extract	Weight loss method and potentiodynamic polarization technique	Langmuir and Temkin adsorption isotherms, mixed-type inhibitor.	48
49	Tin electrode and 0.1 M NaHCO$_3$ solution	Aqueous extract of *Lawsonia*, licorice root, and carob	Galvanostatic polarization measurements, potentiodynamic anodic polarization technique	Freundlich adsorption isotherm.	49
50	Steel and 1 M HCl solution	Extract of brown alga *Bifurcaria bifurcata*	Weight loss, potentiodynamic, and polarization resistance measurements. UV-visible spectrophotometry	Mixed-type inhibitor; formation of a complex at the electrode surface.	50

S. no.	Metal/Medium	Inhibitor/Additive	Method	Findings	Refs.
51	Aluminum and sodium hydroxide	Seed extract of *Abrus precatorius*	Conventional weight loss and polarization measurements techniques.	Good corrosion inhibitor.	51
52	Copper and saline water	Alcoholic extracts of leaves of *Medicago sativa*, *Withania somnifera*, *Atropa belladonna*, and *Medicago polymorpha* and sulfide	EIS, potentiodynamic polarization method, EDS	The complex formation of inhibitor molecules with copper.	52
53	SX 316 steel and HCl solution	Khillah seeds extract	Weight loss measurements and potentiostatic technique	The formation of insoluble complexes.	53
54	C-steel, nickel, zinc, and acidic, neutral, and alkaline solutions	Aqueous extract of the leaves of henna (*Lawsonia inermis*)	Polarization technique.	Mixed-type inhibitor, Langmuir adsorption isotherm.	54
55	Mild steel and HCl (5%)	Acid extract of onion (*Allium cepa*) garlic (*Allium sativum*), and bitter gourd (*Momordica charantia*)	DC electrochemical techniques, classical weight loss method	Freundlich adsorption isotherm, chemisorption.	55

(Continued)

Table 12.1 (*Continued*)

S. no.	Metal/Medium	Inhibitor/Additive	Method	Findings	Refs.
56	Aluminum and 2.0 M HCl solution	Mucilage extracted from the modified stems of prickly pears	Weight loss, thermometry, hydrogen evolution, and polarization techniques	Langmuir adsorption isotherm.	56
57	Mild steel and fresh water	Aqueous extract of *Azadirachta indica* (neem)	Weight loss measurements, potentiodynamic polarization, and AC impedance measurements	Good corrosion inhibitor.	57
58	Steel and commercial aluminum and saline, acidic, and alkaline waters	Water extracts of henna (*Lawsonia inermis*) leaf powder	Weight loss tests	Chemisorption.	58
59	Mild steel and cooling system	Aqueous extracts of *Cordia latifolia* and curcumin and hydroxyethylidene 1-1 diphosphonic acid	Blow down of the cooling system	Good corrosion inhibitor.	59
60	Aluminum and rain water	Garlic extract	Weight loss tests	Good corrosion inhibitor.	60

S. no.	Metal/Medium	Inhibitor/Additive	Method	Findings	Refs.
61	Carbon steel and low-chloride media	Aqueous extract of *Hibiscus rosasinensis* Linn.	Weight loss measurements, potentiodynamic polarization, and AC impedance measurements	The inhibition efficiency increases with increase in extract concentration.	61
62	Mild steel and seawater	*Daucus carota*	Weight loss measurements, potentiodynamic polarization, and AC impedance measurements	Good corrosion inhibitor.	62
63	Carbon steel and seawater	Extract of banana peel	Electrochemical techniques, weight loss method	The inhibition efficiency increases with increase in extract concentration.	63
64	Carbon steel and well water	Caffeine-Zn^{2+}	Electrochemical techniques, weight loss method	The inhibition efficiency increases with increase in extract concentration.	64
65	Carbon steel and seawater	Aqueous extract of henna (*Lawsonia inermis*) leaves	Weight loss measurements, potentiodynamic polarization, and AC impedance measurements	The inhibition efficiency increases with increase in extract concentration.	65

(Continued)

Table 12.1 (*Continued*)

S. no.	Metal/Medium	Inhibitor/Additive	Method	Findings	Refs.
66	Mild steel and seawater	*Cuminum cyminum*	Weight loss measurements, potentiodynamic polarization, and AC impedance measurements	The inhibition efficiency increases with increase in extract concentration.	66
67	Carbon steel and 60 ppm of Cl⁻ ions	Caffeine	Weight loss method, FTIR spectra	A protective film is formed on the metal surface.	67
68	Carbon steel and neutral aqueous solution	Caffeine	Weight loss, fluorescence, FTIR	The inhibition efficiency increases with increase in extract concentration.	68
69	Carbon steel and 60 ppm of Cl⁻ ions	Rhizome (*Curcuma Longa* L.)	FTIR, UV fluorescence, electrochemical studies	Good inhibition efficiency at extreme pH values.	69
70	Carbon steel and well water	Beetroot	Polarization study, AC impedance spectra, FTIR spectra, UV fluorescence	A protective film is formed on the metal surface.	70
71	Carbon steel and aqueous solution containing 60 ppm Cl⁻ ions	Sodium meta vanadate	Weight loss method FTIR spectra, UV-visible spectra, electrochemical studies	A protective film is formed on the metal surface.	71

12.3 Case Study: Inhibition of Corrosion of Mild Steel in Well Water by an Aqueous Extract of *Adhatoda vasica* Leaves

12.3.1 Weight Loss Method

The inhibition efficiency of an aqueous extract of *Adhatoda vasica* leaves in controlling corrosion of mild steel in well water has been evaluated by the weight loss method. The results are summarized in Table 12.2.

Table 12.2 Inhibition efficiency of extract of *A. vasica* leaves in controlling corrosion of mild steel in well water

Volume of *A. vasica* leaf extract (mL)	Corrosion rate (mdd)	Inhibition efficiency (%)
0	32.68	–
2	17.64	46
4	14.05	57
6	9.80	70
8	6.2	81
10	2.61	92

It is observed from the Table 12.2 that as the concentration of *A. vasica* leaf extract increases, the corrosion inhibition efficiency also increases. The active ingredient of *A. vasica* leaf extract, namely vasicine or vasicinone, coordinates with Fe^{2+} ions on the metal surface and forms a protective film consisting of an Fe^{2+}- vasicine or an Fe^{2+}-vasicinone complex. Thus the anodic reaction of metal dissolution is prevented.

12.3.2 Polarization Studies

Polarization studies are useful in confirming the formation of a protective film on the metal surface. If a protective film is formed, the linear polarization resistance increases and the corrosion current value decreases. The polarization curves of mild steel immersed in various test solutions are shown in Fig. 12.1.

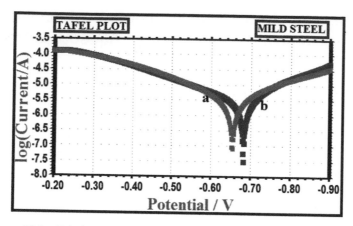

Figure 12.1 Polarization curves of mild steel immersed in various test solutions: (a) well water and (b) well water + 10 mL of *A. vasica* leaf extract.

The corrosion parameters such as corrosion potential (E_{corr}), corrosion current (I_{corr}), linear polarization resistance (LPR), and Tafel slopes (b_c = cathodic; b_a = anodic) are given in Table 12.3.

Table 12.3 Corrosion parameters of mild steel immersed in well water in the absence and presence of an aqueous extract of *A. vasica* leaves (polarization study)

System	E_{corr} (mV vs. SCE)	I_{corr} (A/ cm^2)	LPR (ohm·cm^2)	b_c (mV/ decade)	b_a (mV/ decade)
Well water	−654	9.447×10^{-6}	4022.1	181.6	168.4
Well water + A. vasica leaf extract	−682	3.61×10^{-6}	11108.3	155.9	225.8

It is observed from Table 12.3 that when mild steel is immersed in well water, the corrosion potential is −654 mV versus the saturated calomel electrode (SCE). The corrosion current is 9.447×10^{-6} A/cm^2. The LPR is 4022.1 ohm·cm^2. In the presence of an inhibitor, the corrosion potential shifts from −654 to −682 mV versus the SCE. This is a cathodic shift. It suggests that the cathodic reaction is controlled predominantly. The shift is only in a small range. Hence anodic and cathodic reactions are controlled to an equal extent. The LPR increases from 4022.1 to 11108.3 ohm·cm^2. The corrosion current decreases from 9.447×10^{-6} to 3.61×10^{-6} A/cm^2. These observations confirm that a protective film is formed on the metal surface. This controls the corrosion of metal.

12.3.3   AC Impedance Spectra

AC impedance spectra are useful in confirming the formation of a protective film on the metal surface. If a protective film is formed on the metal surface, the charge transfer resistance (R_t) increases, double-layer capacitance (C_{dl}) decreases, and impedance [log (z/ohm)] increases. The AC impedance spectra of mild steel immersed in well water in the presence of an inhibitor system are shown in the Figs. 12.2 and 12.3. The Nyquist plots are shown in the Fig. 12.2. The Bode plots are shown in Figs. 12.3a and 12.3b. The corrosion parameters, namely R_t, C_{dl}, and impedance, are given in Table 12.4.

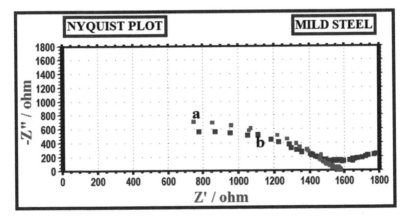

Figure 12.2   AC impedance spectra of mild steel immersed in various test solution (Nyquist plots): (a) well water and (b) well water + 10 mL of *A. vasica* leaf extract.

It is observed from Table 12.4 that when mild steel is immersed in well water, the charge transfer resistance is 286.7 ohm·cm^2, double-layer capacitance is 17.739×10^{-9} F·cm^{-2}, and impedance is 2.721. In the presence of an inhibitor (10 mL of *A. vasica* leaf extract), the charge transfer resistance (R_t) increases from 286.7 to 921 ohm·cm^2, double-layer capacitance (C_{dl}) decreases from 17.739×10^{-9} to 5.428×10^{-9} F·cm^{-2} and impedance increases from 2.721 to 3.201.

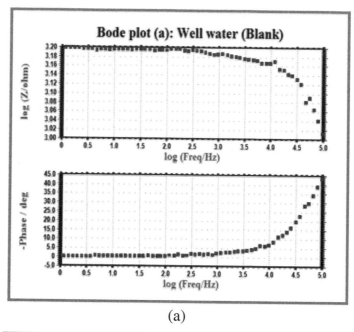

(a)

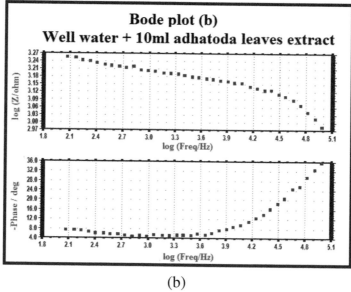

(b)

Figure 12.3 AC impedance spectra of mild steel immersed in well water (Bode plots): (a) well water and (b) well water + 10 mL of *A. vasica* leaf extract.

Table 12.4 Corrosion parameters of mild steel immersed in well water in the presence of *A. vasica* leaf extract obtained from AC impedance spectra

	Nyquist plot		Bode plot
System	R_t (ohm·cm²)	C_{dl} (F·cm⁻²)	Impedance [log(z/ohm)]
Well water (blank)	286.7	17.739×10^{-9}	2.721
Well water + 10 mL of *A. vasica* leaf extract	921	5.428×10^{-9}	3.201

These observations confirm that a protective film is formed on the metal surface. This prevents the transfer of electrons from the metal to the solution medium. Thus corrosion of mild steel is prevented.

12.3.4 AFM Studies

The 2D AFM images of a polished metal surface, corroded surface (immersed in well water), and the film-protected metal (well water + inhibitor solution) are shown in Figs. 12.4a, 12.4b, and 12.4c. The corresponding 3D images are shown in Figs. 12.5a, 12.5b, and 12.5c. The AFM parameters, root-mean-square (RMS) roughness (R_q; nm), average roughness (R_a; nm), and maximum peak-to-valley height (nm) were calculated. These values are given in Table 12.5 for polished mild steel, mild steel immersed in well water, and mild steel immersed in an inhibitor system.

Table 12.5 AFM parameters of mild steel surface in the presence and absence of an inhibitor (*A. vasica* leaf extract)

Sample	RMS roughness (R_q; nm)	Average roughness (R_a; nm)	Maximum peak-to-valley height (nm)
Polished metal	135.9	172.84	806.9
Mild steel immersed in well water (blank)	513.53	637.27	2885.3
Mild steel immersed in well water and 10 mL of *A. vasica* leaf extract	203.86	265.82	1152.6

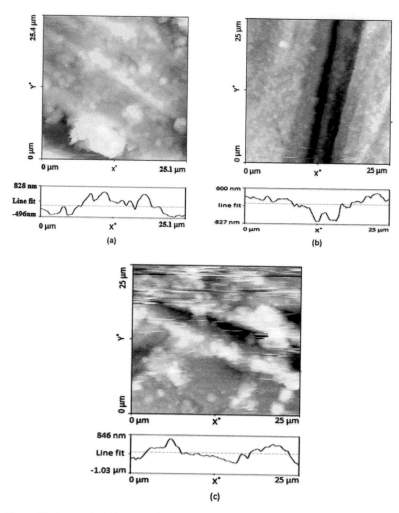

Figure 12.4 2D AFM images of the surface of (a) polished metal, (b) mild steel immersed in well water, and (c) mild steel immersed in well water containing *A. vasica* leaf extract.

It is observed from Table 12.5 that the RMS roughness of polished mild steel is 135.9 nm, average roughness is 172.84 nm, and maximum peak-to-valley height is 806.9 nm. Analysis of Table 12.5 also reveals that the RMS roughness of mild steel immersed in well water is higher than that of polished steel. This indicates that due to corrosion, a thick nanofilm is formed on the metal surface.

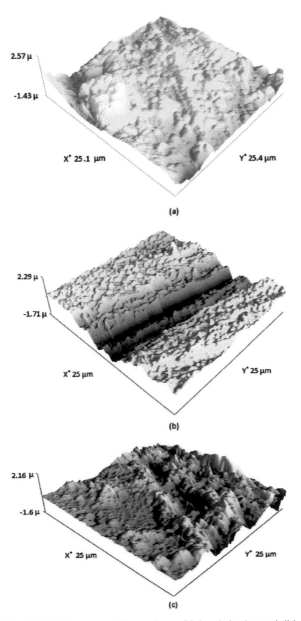

2.57 μ

-1.43 μ

X° 25.1 μm   Y° 25.4 μm

(a)

2.29 μ

-1.71 μ

X° 25 μm   Y° 25 μm

(b)

2.16 μ

-1.6 μ

X° 25 μm   Y° 25 μm

(c)

Figure 12.5 3D AFM images of the surface of (a) polished metal, (b) mild steel immersed in well water, and (c) mild steel immersed in well water containing *A. vasica* leaf extract.

Similar is the case of average roughness and maximum peak-to-valley height. For the metal immersed in an inhibitor system, the RMS roughness is higher than that of the polished metal but lower than that of the metal immersed in a corrosive medium. Similar is the case of average roughness and maximum peak-to-valley height. This indicates that a nanoscale protective film is formed on the metal surface. This film protects the metal from corrosion.

References

1. Umoren, S. A., Solomon, M. M., Obot, I. B., and Suleiman, R. K. (2018). Comparative studies on the corrosion inhibition efficacy of ethanolic extracts of date palm leaves and seeds on carbon steel corrosion in 15% HCl solution, *J. Adhes. Sci. Technol.*, **32**(17), pp. 1934–1951.

2. El-Taib Heakal, F., Deyab, M. A., Osman, M. M., and Elkholy, A. E. (2018). Performance of Centaurea cyanus aqueous extract towards corrosion mitigation of carbon steel in saline formation water, *Desalination*, **425**, pp. 111–122.

3. Trindade, R. S., Dos Santos, M. R., Cordeiro, R. F. B., and D'Elia, E. (2017). A study of the gorse aqueous extract as a green corrosion inhibitor for mild steel in HCl aqueous solution, *Green Chem. Lett. Rev.*, **10**(4), pp. 444–454.

4. Parthipan, P., Narenkumar, J., Elumalai, P., et al. (2017). Neem extract as a green inhibitor for microbiologically influenced corrosion of carbon steel API 5LX in a hypersaline environments, *J. Mol. Liq.*, **240**, pp. 121–127.

5. Tezeghdenti, M., Etteyeb, N., Dhouibi, L., Kanoun, O., and Al-Hamri, A. (2017). Natural products as a source of environmentally friendly corrosion inhibitors of mild steel in dilute sulphuric acid: experimental and computational studies, *Prot. Met. Phys. Chem.*, **53**(4), pp. 753–764.

6. Ajeigbe, S. O., Basar, N., Hassan, M. A., and Aziz, M. (2017). Optimization of corrosion inhibition of essential oils of Alpinia galanga on mild steel using response surface methodology, *ARPN J. Eng. Appl. Sci.*, **12**(9), pp. 2763–2771.

7. Abdel Ghany, N. A., Shehata, M. F., Saleh, R. M., and El Hosary, A. A. (2017). Novel corrosion inhibitors for acidizing oil wells, *Mater. Corros.*, **68**(3), pp. 355–360.

8. Pustaj, G., Kapor, F., and Jakovljević, S. (2017). Carbon dioxide corrosion of carbon steel and corrosion inhibition by natural olive

leaf extract, [Die Kohlendioxid-Korrosion von Kohlenstoffstahl und die Korrosionshemmung durch natürlichen Olivenblattextrakt], *Materialwiss. Werkstofftech.*, **48**(2), pp. 122–138.

9. Ajayi, O., Everitt, N., and Voisey, K. T. (2017). Corrosion inhibition of mild steel in 15 wt.% HCl by durum wheat, Eurocorr 2017: The Annual Congress of the European Federation of Corrosion, 20th International Corrosion Congress and Process Safety Congress 2017.

10. Wang, S., Wu, B., Qiu, L., et al. (2017). Inhibition effect of tangerine peel extract on J55 steel in CO_2 -saturated 3.5 wt. % NaCl solution, *Int. J. Electrochem. Sci.*, **12**(12), pp. 11195–11211.

11. Raghavendra, N., and Bhat, J. I. (2016). Natural products for material protection: an interesting and efficacious anticorrosive property of dry arecanut seed extract at electrode (aluminum)–electrolyte (hydrochloric acid) interface, *J. Bio- Tribo-Corros.*, **2**(4), p. 21.

12. Njoku, D. I., Ukaga, I., Ikenna, O. B., et al. (2016). Natural products for materials protection: corrosion protection of aluminium in hydrochloric acid by Kola nitida extract, *J. Mol. Liq.*, **219**, pp. 417–424.

13. Venegas, R., Figueredo, F., Carvallo, G., Molinari, A., and Vera, R. (2016). Evaluation of Eulychnia acida phil. (cactaceae) extracts as corrosion inhibitors for carbon steel in acidic media, *Int. J. Electrochem. Sci.*, **11**(5), pp. 3651–3663.

14. Salhi, A., Bouyanzer, A., El Mounsi, I., et al. (2016). Natural product extract as eco-friendly corrosion inhibitor for mild steel in 1.0 M HCl, *Der Pharmacia Lettre*, **8**(2), pp. 79–89.

15. Al-Khaldi, M. A. (2016). Natural products as corrosion inhibitor for steel in 0.5 M hydrochloric acid solution, *Asian J. Chem.*, **28**(11), pp. 2532–2538.

16. Emerenciano, D. P., Andrade, A. C. C., De Medeiros, M. L., De Moura, M. F. V., and MaCiel, M. A. M. (2016). The effectiveness of copaiba oil loaded into a microemulsion system as a green corrosion inhibitor, in *Corrosion Inhibitors: Principles, Mechanisms and Applications*, pp. 79–104.

17. Yahia, N. T., Ridha, O. M., and Eddine, L. S. (2016). Inhibitive effect of zygophyllum album extract on the sulfuric acid corrosion of mild steel grade api *5l x52*, *Int. J. Curr. Pharm. Rev. Res.*, **7**(4), pp. 193–197.

18. Abdel Rahman, H. H., Seleim, S. M., Hafez, A. M., and Helmy, A. A. (2015). Study of electropolishing inhibition of steel using natural products as a green inhibitor in ortho-phosphoric acid, *Green Chem. Lett. Rev.*, **8**(3–4), pp. 88–94.

19. De Assis, B. V. R., Meira, F. O., Pina, V. G. S. S., et al. (2015). Inhibitory effect of Piper nigrum L. extract on the corrosion of mild steel in acidic media [Efeito inibitório do extrato de Piper nigrum L. sobre a corrosão do aço carbono em meio ácido], *Rev. Virtual Quim.*, **7**(5), pp. 1830–1840.

20. De Barros, I. B., Moscoso, H. Z. L., Custódio, D. L., Veiga, V. F., and Bastos, I. N. (2015). Aniba canelilla as corrosion inhibitor of carbon steel [Casca preciosa (Aniba canelilla) como inibidor de corrosão do aço-carbono], *Rev. Virtual Quim.*, **7**(5), pp. 1743–1755.

21. Souza, T. F., Magalhães, M., Torres, V. V., and D'Elia, E. (2015). Inhibitory action of Ilex paraguariensis extracts on the corrosion of carbon steel in HCl solution, *Int. J. Electrochem. Sci.*, **10**(1), pp. 22–33.

22. El Mounsi, I., Elmsellem, H., Aouniti, A., et al. (2015). Hexane extract of Nigella sativa L as eco-friendly corrosion inhibitor for steel in 1 M HCl medium, *Der Pharma Chem.*, **7**(10), pp. 350–356.

23. Mounsi, E., Elmsellem, H., Aouniti, A., et al. (2015). Anti-corrosive properties of Nigella sativa L extract on mild steel in molar HCl solution, *Der Pharma Chem.*, **7**(6), pp. 64–70.

24. Alsabagh, A. M., Migahed, M. A., Abdelraouf, M., and Khamis, E. A. (2015). Utilization of green tea as environmentally friendly corrosion inhibitor for carbon steel in acidic media, *Int. J. Electrochem. Sci.*, **10**(2), pp. 1855–1872.

25. Prabhu, R. A., Venkatesha, T. V., Praveen, B. M., Chandrappa, K. G., and Abd Hamid, S. B. (2014). Inhibition effect of Azadirachta indica, a natural product, on the corrosion of zinc in hydrochloric acid solution, *Trans. Indian Inst. Met.*, **67**(5), pp. 675–679.

26. Khadraoui, A., Khelifa, A., Boutoumi, H., and Hammouti, B. (2014). Mentha pulegium extract as a natural product for the inhibition of corrosion. Part I: Electrochemical studies, *Nat. Prod. Res.*, **28**(15), pp. 1206–1209.

27. Fayomi, O. S. I., and Popoola, A. P. I. (2014). The inhibitory effect and adsorption mechanism of roasted elaeis guineensis as green inhibitor on the corrosion process of extruded AA6063 Al-Mg-Si alloy in simulated solution, *Silicon*, **6**(2), pp. 137–143.

28. Sultan, A. A., Ateeq, A. A., Khaled, N. I., Taher, M. K., and Khalaf, M. N. (2014). Study of some natural products as eco - friendly corrosion inhibitor for mild steel in 1.0 M HCl solution, *J. Mater. Environ. Sci.*, **5**(2), pp. 498–503.

29. Tassel, A.-C., Doiron, K., Lemarchand, K., Simard, S., and St-Louis, R. (2014). Corrosion inhibition of mild steel in natural seawater by

snow crab (Chionoecetes opilio) peptidic extracts. Marine peptides as corrosion inhibitor [Inhibition de la corrosion de l'acier doux en eau de mer naturelle par des extraits peptidiques de crabe des neiges (Chionoecetes opilio): peptides marins comme inhibiteur de corrosion], *Mater. Tech.*, **102**(6–7), p. 602.

30. Ali, A. I., Megahed, H. E., El-Etre, M. A., and Ismail, M. N. (2014). Zinc corrosion in HCl in the presence of aqueous extract of Achillea fragrantissima, *J. Mater. Environ. Sci.*, **5**(3), pp. 923–930.

31. Tassel, A.-C., Lemarchand, K., Doiron, K., St-Louis, R., and Simard, S. (2013). Corrosion inhibition of mild steel in seawater by marine biopolymers: physicochemical and biological action, *EUROCORR 2013: European Corrosion Congress.*

32. Umoren, S. A., Gasem, Z. M., and Obot, I. B. (2013). Natural products for material protection: inhibition of mild steel corrosion by date palm seed extracts in acidic media, *Ind. Eng. Chem. Res.*, **52**(42), pp. 14855–14865.

33. Zhang, J., Song, Y., Su, H., et al. (2013). Investigation of Diospyros Kaki L.f husk extracts as corrosion inhibitors and bactericide in oil field, *Chem. Cent. J.*, **7**(1), p. 109.

34. Sangeetha, M., Rajendran, S., Sathiyabama, J., and Krishnaveni, A. (2013). Inhibition of corrosion of aluminium and its alloys by extracts of green inhibitors, *Port. Electrochim. Acta*, **31**(1), pp. 41–52.

35. Belkhaouda, M., Bammou, L., Zarrouk, A., et al. (2013). Inhibition of c-steel corrosion in hydrochloric solution with chenopodium ambrorsioides extract, *Int. J. Electrochem. Sci.*, **8**(5), pp. 7425–7436.

36. Nasibi, M., Zaarei, D., Rashed, G., and Ghasemi, E. (2013). Chamomile (matricaria recutita) extract as a corrosion inhibitor for mild steel in hydrochloric acid solution, *Chem. Eng. Commun.*, **200**(3), pp. 367–378.

37. Oguzie, E. E., Oguzie, K. L., Akalezi, C. O., et al. (2013). Natural products for materials protection: corrosion and microbial growth inhibition using Capsicum frutescens biomass extracts, *ACS Sustainable Chem. Eng.*, **1**(2), pp. 214–225.

38. Oguzie, E. E., Adindu, C. B., Enenebeaku, C. K., et al. (2012). Natural products for materials protection: mechanism of corrosion inhibition of mild steel by acid extracts of piper guineense, *J. Phys. Chem. C*, **116**(25), pp. 13603–13615.

39. Afia, L., Salghi, R., Bazzi, E., et al. (2011). Testing natural compounds: Argania spinosa Kernels extract and cosmetic oil as ecofriendly inhibitors for steel corrosion in 1 M HCl, *Int. J. Electrochem. Sci.*, **6**(11), pp. 5918–5939.

40. Ezhilarasi, J. C., Princey, M. J., and Nagarajan, P. (2011). Natural product Cassia auriculata extract as the corrosion inhibitor for commercial mild steel in 1M HCI: Part-III, *Indian J. Environ. Prot.*, **31**(5), pp. 376–380.

41. Ameer, M. A., Fekry, A. M., and Shanab, S. M. (2011). Electrochemical behavior of titanium alloy in 3.5% NaCl containing natural product substances, *Int. J. Electrochem. Sci.*, **6**(5), pp. 1572–1585.

42. Garg, U., Kumpawat, V., Chauhan, R., and Tak, R. K. (2011). Corrosion inhibition of copper by natural occurring plant Acacia Senegal, *J. Indian Chem. Soc.*, **88**(4), pp. 513–519.

43. Rajalakshmi, R., Subashini, S., and Prithiba, A. (2010). Acid extracts of Ervatamia coronaria leaves for corrosion inhibition of mild steel, *Asian J. Chem.*, **22**(7), pp. 5034–5040.

44. Oguzie, E. E. (2010). Evaluation of the inhibitive effect of some plant extracts on the acid corrosion of mild steel, *NACE - International Corrosion Conference Series*.

45. da Rocha, J. C., da Cunha Ponciano Gomes, J. A., and D'Elia, E. (2010). Corrosion inhibition of carbon steel in hydrochloric acid solution by fruit peel aqueous extracts, *Corros. Sci.*, **52**(7), pp. 2341–2348.

46. Deepa Rani, P., and Selvaraj, S. (2010) Inhibitive action of vitis vinifera (GRAPE) on copper and brass in natural sea water environment, *Rasayan J. Chem.*, **3**(3), pp. 473–482.

47. Morris Princey, J., and Nagarajan, P. (2010). Natural product Eclipta prostrata Linn extract as the corrosion inhibitor for commercial mild steel in 1 M HCl, *Indian J. Environ. Prot.*, **30**(4), pp. 319–323.

48. Subhashini, S., Rajalakshmi, R., Prithiba, A., and Mathina, A. (2010). Corrosion mitigating effect of Cyamopsis Tetragonaloba seed extract on mild steel in acid medium, *E-J. Chem.*, **7**(4), pp. 1133–1137.

49. Abdallah, M., El-Etre, A. Y., Abdallah, E., and Eid, S. (2009). Natural occurring substances as corrosion inhibitors for tin in sodium bicarbonate solutions, *J. Korean Chem. Soc.*, **53**(5), pp. 485–490.

50. Abboud, Y., Abourriche, A., Ainane, T., et al. (2009). Corrosion inhibition of carbon steel in acidic media by Bifurcaria bifurcata extract, *Chem. Eng. Commun.*, **196**(7), pp. 788–800.

51. Rajalakshmi, R., Subhashini, S., Nanthini, M., and Srimathi, M. (2009). Inhibiting effect of seed extract of Abrus precatorius on corrosion of aluminium in sodium hydroxide, *Orient. J. Chem.*, **25**(2), pp. 313–318.

52. El-Dahan, H. A. (2006). Natural products as corrosion inhibitors for copper in saline water in presence of sulphide, *Egypt. J. Chem.*, **49**(5), pp. 589–600.

53. El-Etre, A. Y. (2006). Khillah extract as inhibitor for acid corrosion of SX 316 steel, *Appl. Surf. Sci.*, **252**(24), pp. 8521–8525.

54. El-Etre, A. Y., Abdallah, M., and El-Tantawy, Z. E. (2005). Corrosion inhibition of some metals using lawsonia extract, *Corros. Sci.*, **47**(2), pp. 385–395.

55. Parikh, K. S., and Joshi, K. J. (2004). Natural compounds onion (Allium Cepa), Garlic (Allium Sativum) and bitter gourd (Momordica Charantia) as corrosion inhibitors for mild steel in hydrochloric acid, *Trans. SAEST*, **39**(1–2), pp. 29–35.

56. El-Etre, A. Y. (2003). Inhibition of aluminum corrosion using Opuntia extract, *Corros. Sci.*, **45**(11), pp. 2485–2495.

57. Mohanan, S., Manimegalai, M., Rajeswari, P., Maruthamuthu, S., and Palaniswamy, N. (2002). Biocidal and inhibitive duality of naturally occurring substance azadiracta indica on mild steel in fresh water, *Bull. Electrochem.*, **18**(8), pp. 337–342.

58. Al-Sehaibani, H. (2000). Evaluation of extracts of Henna leaves as environmentally friendly corrosion inhibitors for metals, *Materialwiss. Werkstofftech.*, **31**(12), pp. 1060–1063.

59. Farooqi, I. H., Hussain, A., Quraishi, M. A., and Saini, P. A. (1999). Study of low cost eco-friendly compounds as corrosion inhibitors for cooling systems, *Anti-Corros. Methods Mater.*, **46**(5), pp. 328–331.

60. Priya, S. L., Chitra, A., Rajendran, S. and Anuradha, K. (2005). Corrosion behaviour of aluminium in rain water containing garlic extract, *Surf. Eng.*, **21**(3), pp. 229–231.

61. Anuradha, K., Vimala, R., Narayanasamy, B., Arockia Selvi, J., and Rajendran, S. (2008). Corrosion inhibition of carbon steel in low chloride media by an aqueous extract of Hibiscus rosa-sinensis Linn, *Chem. Eng. Commun.*, **195**(3), pp. 352–366.

62. Sirbharathy, V., Rajendran, S., and Sathyabama, J. (2011). Inhibition of mild steel corrosion sea water by "DAUCUS CAROTA", *Int. J. Chem. Sci. Technol.*, **1**(3), pp. 108–115.

63. Sangeetha, M., Sathyabama, J., Rajendran, S., and Prabhakar, P. (2012). Eco friendly extract of banana peel as corrosion inhibitor for carbon steel in sea water, *J. Nat. Prod. Plant Resour.*, **2**(5), pp. 601–610.

64. Rajam, K., Rajendran, S. and Nazeera Banu, N. (2013). Effect of caffeine-Zn^{2+} system in preventing corrosion of carbon steel in well water, *J. Chem.*, Article ID 521951 (11 pp).

65. Johnsirani, V., Sathiyabama, J., Rajendran, S., and Suriya Prabha, A. (2012). Inhibitory mechanism of carbon steel corrosion in sea water by an aqueous extract of henna leaves, *ISRN Corros.*, Article ID 574321 (9 pp).

66. Sribharathy, V. and Rajendran, S. (2012). Cuminum cyminum extracts as eco-friendly corrosion inhibitor for mild steel in seawater, *ISRN Corros.*, Article ID 370802 (7 pp).

67. Anthony, N., Malarvizhi, E., Maheshwari, P., Rajendran, S., and Palaniswamy, N. (2004). Corrosion inhibition by caffeine-Mn^{2+} system, *Indian J. Chem. Technol.*, **11**, pp. 346–350.

68. Rajendran, S., Amalraj, A. J., Joice, M. J., Anthony, N., Trivedi, D. C., Sundaravadivelu, M. (2004). Corrosion inhibition by the caffeine – Zn^{2+} system, *Corros. Rev.*, **22**(3), pp. 233–248.

69. Rajendran, S., Shanmugapriya, S., Rajalakshmi, T., Amal Raj, A. J. (2005). Corrosion inhibition by an aqueous extract of rhizome powder, *Corrosion*, **61**(7), pp. 685–692.

70. Arockia Selvi, J., Gangasree, V., and Rajendran, S. (2008). Corrosion inhibition by beet root extract, *Port. Electrochim. Acta*, **27**(1), pp. 1–11.

71. Sirbharathy, V., and Rajendran, S. (2012). Corrosion inhibition by green inhibitor: sodium metavadate–spirulina system, *Chem. Sci. Rev. Lett.*, **1**(1), pp. 25–29.

72. Rajendran, S., Jeyasundari, J., Arockia Selvi, J., Narayanasamy, B., Regis, A.P.P., and Rengan, P. (2009). Corrosion behaviour of aluminium in the presence of an aqueous extract of Hibiscus rosa-sinensis, *Port. Electrochim. Acta*, **27**(2), pp. 153–164.

73. Rajendran, S., Paulraj, J., Rengan, P., Jeyasunbdari, J., and Manivannan, M. (2009). Corrosion behaviour of metals in artificial saliva in presence of spirulina powder, *J. Dent. Oral Hyg.*, **1**(1), pp. 1–8.

74. Rajendran, S., Devi, M. K., Regis, A. P. P., Amalraj, A. J., Jeyasundari, J., and Manivannan, M. (2009). Electroplating using environmental friendly garlic extract: a case study, *Zastita Materijala*, **50**(3), pp. 131–140.

75. Rajendran, S., Agasta, M., Bama Devi, R., Shyamala Devi, B., Rajam, K., and Jayasundari, J. (2009). Corrosion inhibition by an aqueous extract of Henna Leaves (Lawsonia inermis L), *Zastita Materijala*, **50**(2), pp. 77–84.

76. Rajendran, S., Sumithira, P., Shyamal Devi, B., and Jeyasundari, J. (2009). Corrosion inhibition by spirulina, *Zastita Materijala*, **50**(4), pp. 223–226.

77. Rajendran, S., Muthulakshmi, S., Rajeswari, R., and Vijitha, A. (2005). An eco friendly corrosion inhibition for aluminium, *J. Electrochem. Soc.*, **54**(2), pp. 50–52.

78. Antony, N., Benita Sherina, H., and Rajendran, S. (2010). Evaluation of co-inhibition characteristics of caffeine-Zn^{2+}system in preventing carbon steel corrosion, *Int. J. Eng. Sci. Technol.*, **2**(7), pp. 2774–2782.

79. Rajendran, S., Ganga Sri, V., Arockiaselvi, J., and Amalraj, A. J. (2005). Corrosion inhibition by plant extracts: an overview, *Bull. Electrochem.*, **21**(8), pp. 367–377.

80. Rajendran, S., Sumithra, P., Shyamala Devi, B., and Jeyasundari, J. (2009). Corrosion inhibition by spirulina, *Zastita Materijala*, **5**, pp. 223–226.

81. Sangeetha, M., Rajendran, S., Muthu Megala, T. S., and Krishnaveni, K. (2011). Green corrosion inhibitiors: an overview, *Zastita Materijala*, **52**(1), pp. 35–19.

82. Syamala Devi, B., and Rajendran, S. (2011). Influence of garlic extract on the inhibtion efficiency of tirsodium citrate - Zn^{2+} system, *Int. J. Chem. Sci. Technol.*, **1**(1), pp. 79–87.

83. Sangeetha, M., Rajendran, S., Sathiya Bama, J., Krishnaveni, A., Santhy, P., Manimaran, N., and Shyamaladevi, B. (2011). Corrosion inhibition by an Aqueous extract of phyllanthus Amarus, *Port. Electrochim. Acta*, **29**(6), pp. 429–444.

84. Felicita Florence, J., Rajendran, S., and Srinivasan, K. V. (2012). Effect of henna (Lawsonia inermis) extract on electrodeposition of nickel from Watt's bath, *Electroplating Finish.*, **31**(7), pp. 1–4.

Index